Lucent Technologies
Bell Labs Innovations

Streamlined
Life-Cycle Assessment

Thomas E. Graedel
Bell Laboratories, Lucent Technologies

now at School of Forestry and Environmental Studies
Yale University

Prentice Hall
Upper Saddle River, New Jersey 07458

Library of Congress Cataloging-in-Publication Data
Graedel, T. E.
    Streamlined life-cycle assessment / Thomas E. Graedel.
      p. cm.
    Includes bibliographical references and index.
    ISBN 0-13-607425-1
    1. Product life cycle--Environmental aspects.   I. Title.
TS170.5.G73   1998
658.4'08--dc21   98-7301
                                         CIP

Acquisition Editor: Bill Stenquist
Editorial/Production: Kelly Dobbs - Douglas & Gayle
Editor-in-Chief: Marcia Horton
Assistant Vice President of Production and Manufacturing: David W. Riccardi
Managing Editor: Bayani Mendoza de Leon
Full Service/Manufacturing Coordinator: Donna Sullivan
Manufacturing Manager: Trudy Pisciotti
Creative Director: Jayne Conte
Cover Designer: Bruce Kenselaar
Editorial Assistant: Meg Weist
Compositor: Douglas & Gayle

©1998 by Bell Laboratories, Lucent Technologies
Published by Prentice-Hall, Inc.
A Division of Simon & Schuster
Englewood Cliffs, New Jersey 07632

All rights reserved. No part of this book may be
reproduced, in any form or by any means,
without permission in writing from the publisher.

The author and publisher of this book have used their best efforts in preparing this book. These efforts include the development, research, and testing of the theories and programs to determine their effectiveness. The author and publisher make no warranty of any kind, expressed or implied, with regard to these programs or the documentation contained in this book. The author and publisher shall not be liable in any event for incidental or consequential damages in connection with, or arising out of, the furnishing, performance, or use of these programs.

Printed in the United States of America

10 9 8 7 6 5 4 3 2 1

ISBN 0-13-607425-1

Prentice-Hall International (UK) Limited, *London*
Prentice-Hall of Autralia Pty, Limited, *Sydney*
Prentice-Hall Canada Inc., *Toronto*
Prentice-Hall Hispanoamericana, S.A., *Mexico*
Prentice-Hall of India Private Limited, *New Dehli*
Prentice-Hall of Japan, Inc., *Tokyo*
Simon & Schuster Asia Pte. Ltd., *Singapore*
Editora Prentice-Hall do Brasil, Ltda., *Rio de Janeiro*

Contents

1 THE GRAND OBJECTIVES   **1**

    1.1 Activities of a Technological Society 1
    1.2 The Technological Process 2
    1.3 Defining the Grand Objectives 3
    1.4 The Crucial Environmental Concerns 6
    1.5 Targeted Activities of Technological Societies 9
    1.6 Directed Actions for an Industrialized Society 9
    1.7 Multi-Objective Decision Making 13
    1.8 Approaches to Assessment 16
    Further Reading 16
    Exercises 17

2 THE CONCEPT OF LIFE-CYCLE ASSESSMENT   **18**

    2.1 The Life Cycle of Industrial Products 18
    2.2 The LCA Framework 21
    2.3 The Industrial Ecology Flow Cycle 24
    Further Reading 28
    Exercises 28

3 GOAL SETTING, SCOPE DETERMINATION, AND INVENTORY ANALYSIS   **29**

    3.1 Goal Setting and Scope Determination 29
    3.2 Defining Study Boundaries 30
        3.2.1 *Life Stage Boundaries* 30
        3.2.2 *Level of Detail Boundaries* 31
        3.2.3 *The Natural Ecosystem Boundary* 31
        3.2.4 *Boundaries in Space and Time* 32
        3.2.5 *Choosing Boundaries* 32
    3.3 The Functional Unit 34
    3.4 Approaches to Data Acquisition 34
    3.5 Process and Product Budgets 36
    3.6 Allocation 39

3.7	Case Study: Detergent-Grade Surfactants	41
Further Reading		42
Exercises		43

4 IMPACT AND INTERPRETATION ANALYSIS — 45

4.1	LCA Impact Analysis	45
4.2	Linking of Stressors and Impacts	47
4.3	Industrial Prioritization: The Netherlands VNCI System	49
4.4	Industrial Prioritization: The IVL/VOLVO EPS System	52
4.5	Industrial Prioritization: The Netherlands NSAEL Method	58
4.6	Interpretation Analysis	62
	4.6.1 Prioritization Tables	62
	4.6.2 Prioritization Diagrams	64
	4.6.2.1 The Action-Agent Prioritization Diagram	64
	4.6.2.2 The Life-Stage Prioritization Diagram	65
	4.6.3 Aspects of Interpretation	66
Further Reading		66
Exercises		67

5 EVALUATING THE LCA PROCESS — 69

5.1	System Boundary Considerations	69
5.2	Functional Unit Considerations	70
5.3	Data Limitation Considerations	70
5.4	Impact Assessment Considerations	71
	5.4.1 Thresholds and Nonlinearities	71
	5.4.2 Temporal Scales	73
	5.4.3 Spatial Scales	74
	5.4.4 Valuation	75
5.5	Pondering LCA Case Studies	76
	5.5.1 Outdoor Clothing	76
	5.5.2 Polyester Blouses	76
	5.5.3 Grocery Sacks	78
	5.5.4 Hot Drink Machines	78
	5.5.5 Food Products	79
	5.5.6 Computer Workstations	80
	5.5.7 Disposable and Reusable Diapers	83
	5.5.8 Retail Sale of Building Materials	84
	5.5.9 Lithographic Printing Operations	84
	5.5.10 Case Study Summary	84
5.6	Assets and Liabilities of the LCA Process	85
Further Reading		85
Exercises		86

6 THE STREAMLINED LCA PHILOSOPHY — 87

- 6.1 The Assessment Continuum — 87
- 6.2 Preserving Perspective — 88
- 6.3 Alternative SLCA Approaches — 89
    - 6.3.1 *The Migros Concept* — 90
    - 6.3.2 *University of British Columbia's and IBM Corporation's SLCA Approach* — 90
    - 6.3.3 *The Dow Chemical Company Matrix* — 90
    - 6.3.4 *The Monsanto Matrix* — 92
    - 6.3.5 *Motorola's SLCA Approach* — 92
    - 6.3.6 *Battelle's Pollution Prevention Factors Approach* — 93
    - 6.3.7 *Jacobs Engineering's SLCA Approach* — 95
- 6.4 Matrix Calculations — 96
- 6.5 SLCA Assets and Liabilities — 97
- Further Reading — 98
- Exercises — 98

7 PRODUCT ASSESSMENT BY SLCA MATRIX APPROACHES — 99

- 7.1 The Manufacturing Challenge — 99
- 7.2 Matrix Concepts — 100
- 7.3 Target Plots — 102
- 7.4 Assessing Generic Automobiles of Yesterday and Today — 103
- 7.5 Discussion — 108
- Further Reading — 110
- Exercises — 110

8 PROCESS ASSESSMENT BY SLCA MATRIX APPROACHES — 111

- 8.1 Considerations in Process Assessment — 111
- 8.2 Stages in Process Life Cycles — 112
    - 8.2.1 *Resource Provisioning* — 114
    - 8.2.2 *Process Implementation* — 114
    - 8.2.3 *Primary Process Operation* — 114
    - 8.2.4 *Complementary Process Operation* — 115
    - 8.2.5 *Refurbishment, Recycling, Disposal* — 115
    - 8.2.6 *The Rating Matrix* — 116
- 8.3 The Approach to Process Analysis — 116
    - 8.3.1 *The Process Itself* — 116
    - 8.3.2 *The Process Equipment* — 117
    - 8.3.3 *Complementary Processes* — 117
- 8.4 Assessing Generic Automobile Manufacturing Processes — 118
    - 8.4.1 *Primary Process Descriptions* — 118

		8.4.1.1 Sand Casting	118
		8.4.1.2 Die Casting	120
		8.4.1.3 Metal Forming	121
		8.4.1.4 Welding	122
		8.4.1.5 Metal Plating	123
		8.4.1.6 Painting	123
		8.4.2 Complementary Process Descriptions	124
		8.4.2.1 Trimming and Smoothing of Castings	124
		8.4.2.2 Metal Cleaning	124
		8.4.3 The Process Assessment	125
	8.5	Conducting the Assessment and Prioritizing the Recommendations	127
	Further Reading		132
	Exercises		132

9 FACILITY ASSESSMENT BY SLCA MATRIX APPROACHES — 134

	9.1	Goals of Facility Assessment	134
	9.2	Life Stages of Industrial Facilities	135
		Stage 1: Site Development, Facility Development, and Infrastructure	137
		Stage 2a: Principal Business Activity: Products	137
		Stage 2b: Principal Business Activity: Processes	138
		Stage 2c: Facility Operations	138
		Stage 3: Facility Refurbishment, Transfer, and Closure	138
	9.3	Life Stages of Office and Residential Facilities	139
		Stage 1a: Site and Infrastructure Development	139
		Stage 1b: Facility Development	139
		Stage 2a: Facility Operations: Indoors	139
		Stage 2b: Facility Operations: Outdoors	139
		Stage 3: Facility Refurbishment, Transfer, and Closure	139
	9.4	Assessment Approaches for Facilities	139
	9.5	Assessing Generic Automobile Manufacturing Plants	142
	9.6	Discussion	147
	Further Reading		149
	Exercises		149

10 SERVICES ASSESSMENT BY SLCA MATRIX APPROACHES — 150

	10.1	The Services Industry	150
	10.2	Type Alpha Services: The Customer Comes to the Service	151
	10.3	Type Beta Services: The Service Goes to the Customer	151
	10.4	Type Gamma Services: Remote Provisioning	153
	10.5	The Services Assessment Matrix	154
		Stage 1: Site and Service Development	154
		Stage 2: Service Provisioning	154

		Stage 3a: Performing the Service	155
		Stage 3b: Facility Operations	155
		Stage 4: Site and Services Closure	155
	10.6	Contracts for Buildings and Equipment	156
	10.7	Assessments of Generic Services	156
	10.8	The Services Economy	167
	Further Reading		169
	Exercises		169

11 INTEGRATED BUSINESS UNIT, CORPORATE, AND INTERCORPORATE ASSESSMENT — 170

	11.1	Introduction	170
	11.2	Integrated Product Assessments	170
	11.3	Integrated Operations Assessments	173
	11.4	Integrated Intercorporate Assessments	175
	11.5	An Integrated Assessment Perspective	178
	Further Reading		180
	Exercises		180

12 ASSESSMENT OF SOCIETAL INFRASTRUCTURE — 182

	12.1	Infrastructure Systems	182
	12.2	Matrix Approaches for Infrastructure Systems	183
		Stage 1: Site Development	183
		Stage 2a: Materials and Product Delivery	183
		Stage 2b: Infrastructure Manufacture	183
		Stage 3: Infrastructure Use	184
		Stage 4: Refurbishment, Recycling, Disposal	184
	12.3	Assessing the Automotive Infrastructure of Yesterday and Today	185
	12.4	Discussion	190
	Further Reading		190
	Exercises		191

13 UPGRADED STREAMLINING — 193

	13.1	The Upgrading Concept	193
	13.2	Bonus Scoring	193
	13.3	Introducing Localization into SCLA Matrix Techniques	195
	13.4	Weighting the Matrices by Consensus	203
	13.5	Weighting the Matrices by the Grand Objectives	207
	13.6	The Localized, Valuated SLCA	213
	Further Reading		214
	Exercises		214

14 REVERSE LCAS — 216

- 14.1 Introduction — 216
- 14.2 Life-Cycle Assessment of the Generic Washing Machine — 216
  - *14.2.1 The Washing Machine Life Cycle* — 217
  - *14.2.2 Assembly Flow* — 218
- 14.3 Streamlined LCA of the Generic Washing Machine — 220
  - *14.3.1 The Matrix Approach* — 220
  - *14.3.2 SLCA Assessment Results* — 223
- 14.4 Needs-Based Approaches — 223
- 14.5 Discussion — 227
- Further Reading — 228
- Exercises — 229

15 THE ULTIMATE LIFE-CYCLE ASSESSMENT — 230

- 15.1 Breadth of Assessment — 230
- 15.2 Efficiency and Tractability — 230
- 15.3 Alternative Tools — 231
- 15.4 The Industrial Ecologist as Physician — 232
- 15.5 The Ultimate LCA — 233
- Further Reading — 234

A ENVIRONMENTALLY RESPONSIBLE PRODUCT MATRIX: SCORING GUIDELINES AND PROTOCOLS — 235

B ENVIRONMENTALLY RESPONSIBLE PROCESS MATRIX: SCORING GUIDELINES AND PROTOCOLS — 250

C ENVIRONMENTALLY RESPONSIBLE FACILITIES MATRIX: SCORING GUIDELINES AND PROTOCOLS — 265

D ENVIRONMENTALLY RESPONSIBLE SERVICES MATRIX: SCORING GUIDELINES AND PROTOCOLS — 278

E ENVIRONMENTALLY RESPONSIBLE INFRASTRUCTURE MATRIX: SCORING GUIDELINES AND PROTOCOLS — 291

F VALUES OF LOCALIZATION PARAMETERS FOR COMMON STRESSORS — 305

INDEX — 306

Preface

Many factors, including governmental requirements, good citizenship, and (especially) customer demands, encourage industrial corporations to move in the direction of environmental responsibility in their operations. The approaches taken in the past have been largely defined by regulatory agencies and looked upon as a necessary cost of doing business but hardly as a driving force for engineers and managers. The emerging modern view of environmental responsibility is much broader than has historically been the case. It envisions an approach in which one looks not to the past, but to the future, and has as the goal the design and manufacture of products, the operation of processes, and the management of facilities so that environmental factors are recognized and their impacts minimized. Such actions are increasingly viewed as sound business practice in a competitive world concerned both with the perceptions of customers and with the planet on which we live.

These are attractive but fuzzy notions, and the fuzziness can readily be seen when a designer attempts to decide whether a product, process, or system is "green." What makes one design environmentally preferable to another? How can we know whether we are doing what we claim to be doing? The answers to at least some of these questions may be found in the techniques termed *life-cycle assessment* (LCA) and its streamlined variations.

Life-cycle assessments are ultimately rooted in the grand environmental objectives of society, as described in Chapter 1. Following that introduction, LCAs of various kinds form the principal subject of the book. General LCA concepts are explained in Chapter 2. The first step in an LCA is the setting of goals and the determination of the scope of the study. These topics are presented in Chapter 3, which also includes a discussion of data collection and presentation.

Once all of the possible design options are addressed as part of the inventory activity, they must be compared and prioritized. The prioritization depends strongly on the degree to which a given material perturbs one of the natural Earth system budgets or has a detrimental impact on human organisms or systems. Chapter 4 describes various approaches for impact assessment. This LCA stage is where one examines how the actions of industry and society influence the environment. Assessing the degree of that interaction is difficult, complex, and extremely value-laden. Chapter 5 discusses some of these and other problems with LCA as currently practiced.

The LCA process in its full implementation turns out, once carefully examined, to be costly in resources and time, to provide results that are often open to interpretation, and to be essentially unworkable as a routine tool. Nonetheless, the benefits of an LCA framework are widely recognized. As a consequence, corporations have developed a number of techniques for streamlining the LCA process. A number of these are reviewed in Chapter 6. Chapters 7–11 then utilize a widely-used *streamlined life-cycle assessment* (SLCA) approach to describe streamlining in detail and to demonstrate its use in the assessment of products, processes, facilities, and services. Chapter 12 treats a more all-encompassing and societal topic—infrastructures—and asks how those systems may be assessed for their environmental merit. Chapter 13 explores ways in which SLCA techniques may be enhanced by bonus scoring and weighting factors and how they may be applied to reflect the characteristics of the geographic region in which the activites are occurring.

The last two chapters are philosophical in nature. They ask what the environmentally ideal product might look like and what the characteristics of the ultimate SLCA might be. The discussions suggest interesting goals for the future.

I hope this book will have several uses. The first is as a case study in design courses in automotive, mechanical, and civil engineering classes. Second, I hope practicing engineers will find it of value as they continue to improve the environmental performance of their product and process designs. Third, I anticipate that the topic will be of interest to students of business and management, now striving to blend the challenges of the environment with those of industry and the world within which it functions.

Life-cycle assessments are an operational tool that serve a broader approach to the interactions of society and the environment termed *industrial ecology*. Industrial ecology is not the province only of industry and those associated with it, and I mention in this book only in passing a number of related topics that profit from more detailed discussion: technological change, risk assessment, economic and legal implications, corporate structure, governmental interests, and the like. Those interested in more information are referred to the full-length textbook *Industrial Ecology* (T.E. Graedel and B.R. Allenby, Prentice Hall, Inc., Englewood Cliffs, NJ, 1995).

The first draft of this book was read in its entirety by Braden Allenby, Gören Finnveden, William Hoffman, J. William Owens, and Joel Ann Todd. Their comments and suggestions have markedly improved the book, although they are, of course, not responsible for its contents. I thank William Stenquist and Marcia Horton of Prentice Hall for their continued enthusiasm for publishing volumes relevant to industrial ecology.

Thomas E. Graedel
New Haven, Connecticut
August, 1997

CHAPTER 1

The Grand Objectives

1.1 ACTIVITIES OF A TECHNOLOGICAL SOCIETY

Modern living, and all the things that the term "modern" implies, bears little resemblance to the pastoral lifestyle of our ancestors. Instead, we operate in a technological world. From our factories pour products whose sizes range from 500-passenger airplanes to computer chips the size of a fingernail, whose in-service lifetimes range from several decades (as in construction materials) to a single use (as in clinical supplies for medical laboratories), whose constituent materials may be common minerals or molecular composites of stunning complexity.

In this milieu, humanity has begun to recognize the stresses that modern technological activity and the sheer number of people are placing on the planet, and is asking itself how to make each of its activities more environmentally responsible. Which decisions are the best ones? Why? How can they be translated from concepts into implementation? Questions like these can be given answers only through a structured analytical approach to everything we do as a technological society: how we design our products, how we manufacture them, how we use them, and how we dispose of them. But products alone do not constitute the sum of our technological activities. We must look also at the way we build and use our buildings, the way we provide and receive services of all kinds, indeed, the way in which we design our entire human habitat. As a society, this sort of thinking has the potential to break new ground, yet in the rush of everyday duties it has not received very much attention. Nonetheless, the issues must be addressed, lest our lack of foresight lead to difficulties or even crises that could have been avoided with more insightful thinking and planning.

This book describes ways in which decision makers—within corporations and without—can assess in efficient yet comprehensive ways the environmental implications of the activities of our technological society. There is no single, correct way to evaluate the environmentally related attributes of products, processes, facilities, and services, and the techniques that are currently being used are in the midst of rapid and rather chaotic development. Still, great strides forward have been made with the tools that are available. In one way or another, these tools first promote a system-wide examination of the activity under study, next identify possible modifications, and finally promote consideration of tradeoffs among alternatives. If the tools are used conscientiously, the problematic interactions of society with its environment can be very much mitigated. It is fair to say that environmentally oriented actions will become the norm as the society of the 21st century unfolds.

1.2 THE TECHNOLOGICAL PROCESS

Industrial systems operate within societies and their economic structures, rather than distinct from them. This relationship produces benefits, such as the creation and expansion of markets, and it also produces liabilities, such as environmental impacts. As a consequence, industrial systems are constrained by governmental policies and regulations, and, more broadly, by social morays and economic and technological conditions.

All industrial activity is a response to society's needs and wants. The terms "needs and wants" have a variety of meanings, but from the standpoint of the industrial system both generate demand for products. Sample needs and wants of different societal constituencies at different spatial scales are shown schematically in Table 1.1, transportation being used as an illustration. At the local level, the desire of government for development leads to the construction of rail lines and highways, thus allowing producers ready access to markets and labor supplies. Individual consumers use these facilities not only for commuting to work, but also for private tasks such as shopping. In many cases, it is immaterial to the customer exactly how the transportation is provided. If public transportation is sufficiently convenient and inexpensive, it will be used. Otherwise, customers will prefer to operate private automobiles even though ownership is often expensive and adds complexity to daily living.

As spatial scales become greater, planning assumes longer time horizons, the movement of goods and services is a central focus, and individual transportation becomes less and less central. Transportation as a component of security and competitiveness assumes interest at the national scale. At the international scale, factors such as the opening of markets and the provision for shipment of large quantities of manufactured goods become elements of transportation planning. The message of Table 1.1 is that there are many facets of the industry-society relationship and many levels of motivation and constraint. The sum of these forces produces society's needs and wants.

After needs and wants are identified, how does industry act within society to respond to the resulting demands? A concept for such interactions is shown in Figure 1.1. The flow of information in the figure begins with the needs and wants in the upper left of the diagram. These motive forces are modified by various societal factors, economic constraints, concerns regarding hazards and environmental impacts, and the state of technology. The result is a demand for specific goods and services. Industrial corporations, responding in their own ways to information available to them, design, evaluate, and produce these goods and services.

TABLE 1.1 The Needs and Wants of Principal Constituencies for Transportation, Evaluated at Different Spatial Scales

Constituency	Spatial Scale		
	Local	National	International
Government	Regional development	National security	Trade competitiveness
Primary producers	Dedicated systems	Dedicated systems	Market diversity, stability of demand
Secondary producers	Labor supply	Product distribution, market access	Exports, market presence
Consumers	Commuting, shopping	Recreation, business	Vacation, business

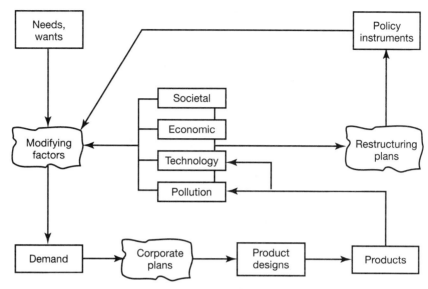

FIGURE 1.1 Interactions between industrial activities and societal systems. The irregular boxes identify planning stages necessary to connect forcing functions and responses.

The interactions shown in the figure demonstrate the role of each individual's decisions in stimulating the activities of corporations, and make it clear that the technological society is not a collection of nameless and far-distant corporations, but rather the sum of the decisions of all of us.

1.3 DEFINING THE GRAND OBJECTIVES

Industrial design and development activities are increasingly guided by the results of *life-cycle assessments* (*LCAs*), which are (1) determinations of associated flows of energy and materials, (2) evaluations of the environmental stresses and impacts chargeable to a product or process from its inception to its end of life, and (3) recommendations designed to improve the environmentally related attributes. The simple act of invoking the idea of a life cycle as part of the technological development process has great value because it sets a framework for whatever analysis is performed. However, while the concept of life-cycle assessment is one that is readily understood and appreciated, its implementation has often proved intractable—or at least impractical—because of problems related to data needs, time, expense, and uncertainty regarding the defendability of the results. This situation has led to the development of the *streamlined LCA* (*SLCA*), which attempts to retain the basic LCA concept while making implementation more efficient and straightforward.

One group of those involved in LCA efforts might be called the *LCA developers*, whose goal is to develop a rigorous and defendable LCA methodology. Although protocols for several common steps in LCAs have indeed been developed, comprehensive LCA has foundered over the difficulties of impact analysis, that is, quantitatively relating a specific product design or a specific process step to the environmental impacts that result. Ideally, a value set would be used to prioritize the various environmental impacts and thus call out some of the LCA recommendations for special treatment. However, no truly satisfactory approaches have yet been devised to take into account such factors as differences in the spatial and temporal scales of impacts, the existence of thresholds, and nonlinear dose-response functions. Furthermore, the operation of determining the environmental impacts of the flows of energy and materials related to a particular technological activity is often more a topic for discussion than a procedure in general use.

Meanwhile, a second group that might be termed the *LCA practitioners* has been constructing practical, if not intellectually rigorous, LCA tools and proceeding to apply them to the assessment of products. This activity has proven successful because there has been an abundance of "low-hanging fruit," that is, actions with high benefit/cost ratios (such as pollution reduction and dematerialization) that are so obviously advantageous to the environment that they raise few ethical quandaries.

What is the current perspective of the LCA developers and the LCA practitioners? In the case of the developers, the lack of agreement on how to construct a value set has led to stasis, and even to suggestions that valuation is inappropriate for LCA. Help is needed to surmount this barrier. In the case of the practitioners, they are finding that much of the low-hanging fruit has been picked, and that efforts to find and harvest more fruit bring them face to face with ethical questions for which their previous approach is not adequate. For example, is it preferable in a manufacturing operation to use a volatile organic solvent or to utilize more energy to do the same job with an aqueous solvent? Is it important to use recycled materials, and, if so, why? In the midst of these complexities, the practitioners remain committed to the concept of the production of green products by green processes, and are in need of guidance on how to proceed.

Efforts to generate practical and reliable LCAs and SLCAs have now been underway for several years. The ultimate aim of the exercises is to produce recommendations to decision makers (product designers, manufacturers, policy makers, and so on) that are intended to minimize the environmentally related attributes of technological activities over their entire life span. In theory, the recommendations can be related to the amelioration of clearly defined and generally accepted environmental concerns. In practice, the relationships between the recommendations and the environmental stresses are often far from straightforward, and may incorporate assumptions not necessarily consistent with the best and most recent perspectives of environmental science. In order to establish firmly the technology-environmental science link, it is necessary to provide an intellectual framework from which the recommendations of LCAs and SLCAs can be justified.

In a general sense, it is clear that guidance on decisions that are not immediately obvious must be based on some sort of ethical judgment. Ethical judgments in a pluralistic society, particularly a global one, are far from uniform. Nonetheless, there has been remarkable agreement on a unifying vision for the interactions between humanity and the environment: that of sustainable development. The Brundtland Commission (the 1987 World Commission on Environment and Development) defined sustainable development as comprising those activities that "meet the needs of the present without compromising the ability of future generations to meet their own needs." Because of the international standing of the Commission and the overwhelming acceptance of the thrust of its report, this vision can already be regarded as something of a societal consensus statement (if perhaps a rather vague one). In addition, there is indisputable evidence that some environmental concerns are regarded generally or even universally as more important than others. For example, a major global decrease in biodiversity is clearly of more concern than the emission of volatile constituents of drying paint, and the Montreal Protocol and the Rio Treaty demonstrate that at least most of the countries of the world feel that understanding and minimizing the prospects for ozone depletion and global climate change are issues of universal importance.

If one accepts that there are indeed such issues that have general acceptance by the human society, one may then postulate the existence of a small number of "grand objectives" having to do with life on Earth, its maintenance, and its enjoyment. A reasonable exposition of society's grand objectives is the following:

- The Ω_1 Objective: Maintaining the existence of the human species.
- The Ω_2 Objective: Maintaining the capacity for sustainable development.
- The Ω_3 Objective: Maintaining the diversity of living things.
- The Ω_4 Objective: Maintaining the aesthetic richness of the planet.

If it is granted that these objectives are universal, it follows that there are certain basic societal requirements that must be satisfied if the objectives are to be met. In the case of Ω_1, these are the minimization of environmental risk and the provision of basic needs: food, water, shelter. For Ω_2, the requirements are a dependable energy supply and the availability of suitable material resources. For Ω_3, it is necessary to maintain a suitable amount of natural areas and to maximize biological diversity on disturbed areas, through the avoidance of large-area monocultural vegetation, for example. Ω_4 requires control of residues of various kinds: minimizing emissions that result in smog, discouraging activities that lead to degradation of the visible world, encouraging farming and agricultural practices that avoid land overuse and erosion, and the preservation of commonly held undeveloped land. The relationships are summarized in Table 1.2.

A potential objection to this approach is the difficulty of reaching societal consensus, either because of a clash of interest around environmental matters, because of ideological or cultural differences, or because of the complexities of organization and collective decision making. Whether or not complete consensus is achieved, the framework developed here can be viewed as a benchmark against which decisions about the use and goals of LCA can be made. Articulation of the grand objectives and their implications for environmental decision making can thereby push product designers, environmental analysts, and others to relate their choices explicitly to larger environmental concerns even without agreement on the relative importance of those concerns.

The grand objectives framework is an important prerequisite to determining what societal activities would be desirable, but the framework does not ensure progress toward achieving the objectives, nor does it deal with the thorny problem of tradeoffs among the omega objectives themselves. Nonetheless, progress results when desirable actions encouraged by some framework of this type occur over and over again. In an industrialized society, many (if not most) of those actions are decisions made by product designers and manufacturing engineers in response to the needs or

TABLE 1.2 Relating Environmental Concerns to the Grand Objectives

Grand Objective	Environmental Concern
Ω_1: Human species extinction	1. Global climate change
	4. Human organism damage
	5. Water availability and quality
	6. Resource depletion: fossil fuels
Ω_2: Sustainable development	5. Water availability and quality
	6. Resource depletion: fossil fuels
	7. Soil depletion
	8. Optimal land use
	12. Resource depletion: other than fossil fuels or soils
Ω_3: Biodiversity	1. Global climate change
	2. Loss of biodiversity
	3. Stratospheric ozone depletion
	5. Water availability and quality
	7. Acid deposition
	16. Thermal pollution
Ω_4: Aesthetic richness	10. Smog
	11. Aesthetic degradation
	13. Oil spills
	15. Odor

The order of the numbers in the right column is that of Table 1.3.

desires of society. Thus, one can envision that recommendations arising from LCAs and SLCAs, if informed by the grand objectives, will become the means by which favorable decisions can be made.

1.4 THE CRUCIAL ENVIRONMENTAL CONCERNS

Grand objectives are, of course, too general to provide direct guidance to the LCA practitioner, who deals with specific actions relating to environmental concerns. Objectives and concerns can readily be related (refer to Table 1.2), but LCA decisions often require, in addition, a ranking of the relative importance of those concerns. This requirement is, in fact, a throwback to the utilitarian philosophy that societal actions should be taken so as to produce the maximization of the good, and produces in turn the question "How does society identify the best actions?"

The particular difficulty in identifying the best actions in this instance is that societal activities related to the environment inevitably involve tradeoffs: wetland preservation versus job creation, the lack of greenhouse gas emissions of nuclear power reactors versus the potential of nuclear accident, or the preservation and reuse of clothing versus the energy costs required for cleaning, to name but a few. To enable choices to be made, many have proposed that environmental resources (raw materials, plant species, the oceans, and so forth) be assigned economic value so that decisions could be market-driven. The concept, though intellectually attractive, has proven difficult to reduce to practice, and has been further confounded by the fact that the scientific understanding of many of the issues to be valued is itself evolving and thus requires that valuation be continuously performed if the analytical framework is to keep pace with the science.

Given this uncertain and shifting foundation for relative ranking, how might specific environmental concerns, several of which might be responsive to one or more of the grand objectives, be grouped and prioritized in an organized manner? One can begin by taking the guidance of the World Commission on Environment and Development that sustainability ultimately requires

- Not using renewable resources faster than they are replenished.
- Not using non-renewable resources faster than renewable substitutes can be found for them.
- Not significantly depleting the diversity of life on the planet.
- Not releasing pollutants faster than the planet can assimilate them harmlessly.

Similarly, The Natural Step, a Swedish organization now expanding into other countries, has offered the "system conditions":

- Substances from Earth's crust must not systematically increase in nature.
- Substances produced by society must not systematically increase in nature.
- The physical basis for the productivity and diversity of nature must not be systematically diminished.
- Make fair and efficient use of resources with respect to meeting basic human needs.

Given either set of guidelines, or any other acceptable set, the relative significance of specific environmental impacts can then be established by consideration of those guidelines in accordance with several related characteristics:

- The spatial scale of the impact (large scales being worse than small).
- The severity of the hazard, i.e., the product of the damage potential of a material, how much material is involved, and how numerous is the exposed population (highly hazardous substances being of more concern than less highly hazardous substances).
- The degree of exposure (well-sequestered substances being of less concern than readily-mobilized substances).
- The penalty for being wrong (longer remediation or reversibility times being of more concern than shorter times).

These characteristics are perhaps too anthropocentric as stated, but are nonetheless a reasonable starting point for distinguishing highly important concerns from those that are less important. Using the criteria and the grand objectives, common local, regional, and global environmental concerns can be grouped as shown in Table 1.3. The exact wording and relative positioning of these concerns is not critical for the present purpose; what is important is that all actions of industrial society that have potentially significant environmental implications relate in some way to the list.

TABLE 1.3 Significant Environmental Concerns

Crucial Environmental Concerns

1. Global climate change
2. Loss of biodiversity
3. Stratospheric ozone depletion
4. Human organism damage
5. Water availability and quality
6. Depletion of fossil fuel resources

Highly Important Environmental Concerns

7. Soil depletion
8. Suboptimal land use
9. Acid deposition
10. Smog
11. Aesthetic degradation
12. Depletion of resources other than fossil fuels

Less Important Environmental Concerns

13. Oil spills
14. Radionuclides
15. Odor
16. Thermal pollution
17. Landfill exhaustion

The numbers within the groupings are for reference purposes, and do not indicate order of importance.

Of the six crucial environmental concerns, three are global in scope and have very long time scales for amelioration: global climate change, loss of biodiversity, and ozone depletion. The fourth critical concern relates to damage to the human organism by toxic, carcinogenic, or mutagenic agents. The fifth critical concern is the availability and quality of water, a concern that embraces the magnitude of water use as well as discharges of harmful residues to surface or ocean waters. The sixth is the rate of loss of fossil fuel resources, which will be vital to many human activities over the next century, at least. The last three also introduce consideration of spatial inhomogeneity because they play out differently in different geographical locations.

Six additional concerns are regarded as highly important, but not as crucial as the first six. The first two of these, soil depletion and suboptimal land use, have to do with the local loss of soils and habitats. The next two, acid deposition and (in some instances) smog, are regional scale impacts occurring in many parts of the world and closely related to fossil fuel combustion and other industrial activities. Aesthetic degradation, the fifth highly important concern, incorporates quality of life issues such as visibility, the action of airborne corrodants on statuary and buildings, and the dispersal of solid and liquid residues. The final concern, depletion of non-fossil fuel resources, is one of the motivations for current efforts to recycle materials and minimize their use.

Finally, five concerns are rated as less important than those in the first two groupings, but still worthy of being called out for attention: oil spills, radionuclides, odor, thermal pollution, and depletion of landfill space. The justification for inclusion in this grouping is that the effects, while sometimes quite serious, tend to be local or of short time duration, or both, when compared with the concerns in the first two groups.

It is appropriate to comment briefly on the selections of the crucial concerns. Of those six, the first four are called out by the U.S. EPA Science Advisory Board and by most detailed approaches to life-cycle assessment. The availability and quality of water is already a major concern in some areas of the world, and will likely become more so as populations and urbanization continue to increase over the next half-century. The concern for fossil fuel resources is obvious, as most of the energy-consuming activities of modern society are dependent on fossil fuels and their supporting infrastructures, and long transition periods will be required to implement any broad replacement of fossil fuels by other energy sources.

1.5 TARGETED ACTIVITIES OF TECHNOLOGICAL SOCIETIES

The mitigation of the environmental impacts of human activities follows, at least in principal, a logical sequence. First is the recognition of an environmental concern related to one or more of the grand objectives. Global climate change, for example, is related to two: Ω_1 and Ω_3. Next, after that concern is identified, environmental scientists study the activities of humanity that are related to it (an initial list is provided in Table 1.4). For global climate change, the activities for examination would include (though not be limited to) those that result in emissions of greenhouse gases, especially CO_2, CH_4, N_2O, and CFCs. Figure 1.2 illustrates the relationship schematically. The concept is that societal activities from agriculture to manufacturing to transportation can be evaluated with respect to their impacts on the grand objectives, and that the link between activities and objectives is the purpose for the environmental evaluation of products, processes, and facilities.

1.6 DIRECTED ACTIONS FOR AN INDUSTRIALIZED SOCIETY

The final step in the structured life-cycle assessment process is that given activities for examination, the life-cycle analyst can generate "design for environment" recommendations, as shown in Figure 1.3, for energy use activities related to global climate change. These recommendations are for specific actions that can be taken by industrial design and manufacturing engineers to improve the environmental responsibility of their products.

Thus, the overall LCA process described here occurs in four stages: (1) the definition by society of its grand environmental objectives for life on Earth, (2) the identification by environmental scientists and policy makers of environmental concerns related to one or more of those objectives, (3) the identification by environmental scientists of the activities of society with the potential to influence those concerns, and (4) the appropriate modification of engineering and societal activities. If one looks carefully at this sequence, notice that implementing the fourth step depends on accepting

TABLE 1.4 Targeted Activities in Connection with Environmental Concerns

1. Global climate change	1.1 Fossil fuel combustion (CO_2 emission)
	1.2 Cement manufacture (CO_2 emission)
	1.3 Rice cultivation (CH_4 emission)
	1.4 Coal mining (CH_4 emission)
	1.5 Ruminant populations (CH_4 emission)
	1.6 Waste treatment (CH_4 emission)
	1.7 Biomass burning (CO_2, CH_4 emission)
	1.8 Emission of CFCs, HFCs, N_2O
2. Loss of biodiversity	2.1 Loss of habitat
	2.2 Fragmentation of habitat
	2.3 Herbicide, pesticide use
	2.4 Discharge of hazardous chemicals to surface waters
	2.5 Reduction of dissolved oxygen in surface waters
	2.6 Oil spills
	2.7 Depletion of water resources
	2.8 Industrial development in fragile ecosystems
3. Stratospheric ozone depletion	3.1 Emission of CFCs
	3.2 Emission of HCFCs
	3.3 Emission of halons
	3.4 Emission of nitrous oxide
4. Human organism damage	4.1 Emission of hazardous materials to air
	4.2 Emission of hazardous materials to water
	4.3 Disposition of hazardous materials in landfills
	4.4 Depletion of water resources
	4.5 Physical organism damage
5. Water availability and quality	5.1 Consumptive use of surface water
	5.2 Use of herbicides and pesticides
	5.3 Use of agricultural fertilizers
	5.4 Discharge of hazardous materials to surface or ground waters
	5.5 Siltation and salinization of surface or ground waters
	5.6 Depletion of water resources
6. Resource depletion: fossil fuels	6.1 Use of fossil fuels for energy
	6.2 Use of fossil fuels as feedstocks
7. Soil depletion	7.1 Soil erosion
	7.2 Discarding or depositing trace metals onto soil
	7.3 Loss of arable land to development
8. Suboptimal land use	8.1 Loss of arable land to development
	8.2 Habitat destruction
	8.3 Abandonment of developed land
9. Acid deposition	9.1 Fossil fuel combustion
	9.2 Emission of sulfur oxides to air
	9.3 Emission of nitrogen oxides to air
10. Smog	10.1 Fossil fuel combustion
	10.2 Emission of VOCs to air
	10.3 Emission of nitrogen oxides to air
11. Aesthetic degradation	11.1 Emission of particulate matter to air
	11.2 Emission of sulfur oxides to air
	11.3 Incomplete combustion of fossil fuels
	11.4 Biomass burning
	11.5 Loss of habitat
	11.6 Oil spills
	11.7 Discarding solid residues
	11.8 Discarding liquid residues

Section 1.6 Directed Actions for an Industrialized Society

TABLE 1.4 Continued.

12. Resource depletion other than fossil fuels and soils	12.1 Use of metals in limited supply
	12.2 Habitat destruction
	12.3 Use of biomaterials
	12.4 Discarding solid residues
	12.5 Discarding liquid residues
	12.6 Discarding gaseous residues
13. Oil spills	13.1 Transport of petroleum
	13.2 Refining of petroleum
	13.3 Distribution of petroleum products
14. Radionuclides	14.1 Production of nuclear power
	14.2 Manufacture of products containing radioisotopes
15. Odor	15.1 Odorous industrial emissions
	15.2 Untreated odorous residues
16. Thermal pollution	16.1 Discharge of heated water to surface waters
	16.2 Discharge of heated water to groundwaters
	16.3 Discharge of heated air
17. Landfill Exhaustion	17.1 Disposition of solid residues in landfills
	17.2 Disposition of liquid residues in landfills

Relating Targeted Activities To Environmental Concerns

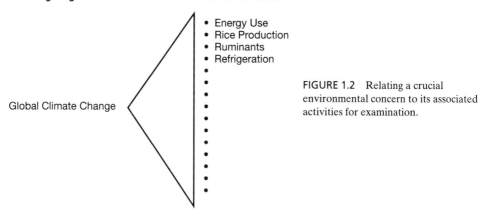

FIGURE 1.2 Relating a crucial environmental concern to its associated activities for examination.

the definition of step one, believing the validity of step two, and acknowledging the correct attribution in step three, but not necessarily in knowing the magnitude of the impact of step four on improving the environment, that is, the information that is needed tends to be qualitative, not quantitative. From the standpoint of the industrial manager or the product design engineer, what is important is knowing that if a step four action is taken, an environmentally disadvantageous situation will be ameliorated to at least some degree, and, perhaps at least as important, knowing that customers and policy makers will regard the action as a positive and thoughtful one.

In overview, the four steps of the process are schematically displayed in Figure 1.4, in which it is clear that the grand objectives determine the importance of each of the environmental concerns, and that each of the environmental concerns leads to a

Relating Specific Recommendations To Targeted Activities

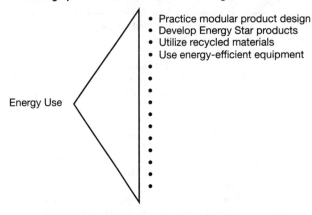

FIGURE 1.3 Relating an activity for examination to associated recommendations for ameliorating its impact on a crucial environmental concern.

The Conceptual Sequence in Life-Cycle Assessment

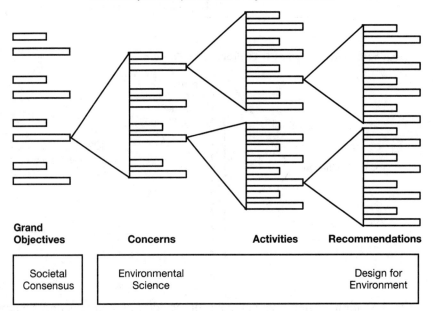

FIGURE 1.4 The conceptual sequence in life-cycle assessment. The manner in which societal consensus, environmental science, and industrial ecology are linked from this perspective is indicated at the bottom of the diagram.

group of activities for examination, each of which in turn leads to a set of product design recommendations. As shown at the bottom of the diagram, the relationships among the grand objectives and recommendations provide logical interconnections among societal consensus, environmental science, and technological actions.

The sequence of Figure 1.4 should not be taken to imply that all actions that society wants to take are encompassed by the environmental grand objectives. Society has many other objectives, some economic, some political, some familial. A framework such as that which the grand objectives provides is not intended to suggest that environmental considerations should be paramount in all human activities, but it does raise those considerations as important ones to be weighed in balancing the often-conflicting desires of our modern technological society.

With a grand objectives framework as a guide to environmentally desirable actions, forward movement requires no further intellectual justification of "design for environment" activities, but rather the development and evaluation of tools to guide assessments and corrective actions.

1.7 MULTI-OBJECTIVE DECISION MAKING

While the sequence of relationships illustrated in Figure 1.4 enables one to justify a particular action from an environmental standpoint, a complication is that most technological activities influence more than one environmental concern (Fig. 1.5) and most concerns are related to more than one technological activity (Fig. 1.6). The goal of industrial ecology decision making then changes from having the most favorable interaction with a single concern to having the optimum set of interactions with a group of concerns. The stakes are now not merely the identification of actions guided by the related grand objectives, but the prioritization of those actions.

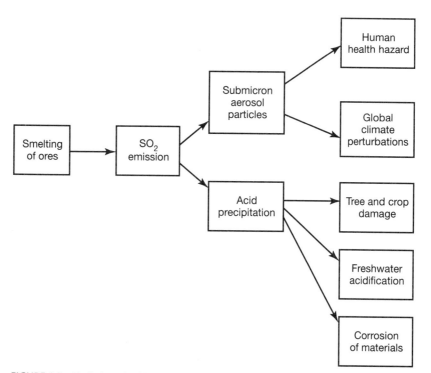

FIGURE 1.5 Emission of sulfur dioxide from smelting of ores, an example of multiple environmental impacts related to a single technological activity.

14 Chapter 1 The Grand Objectives

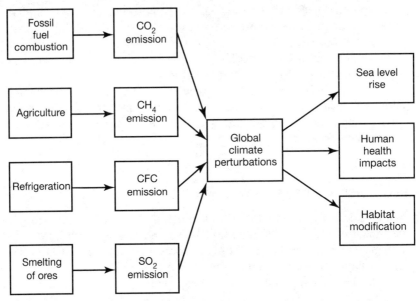

FIGURE 1.6 Global climate perturbations, an example of a single environmental concern related to multiple technological activities.

This multi-objective situation can be considered with the help of an illustration devised by Frank Field and his associates at the Massachusetts Institute of Technology. In this illustration, there are seven possible ways of accomplishing a desired activity; they accomplish it equivalently in all ways except for the environmental interactions. All the ways have impacts on environmental concerns A and B, the values of the impacts being unique for each choice (Fig. 1.7a). Which choice is environmentally preferable?

One can immediately discard choices 2 and 3, as other alternatives obviously have less impact on each of the environmental concerns. It is less straightforward to choose from the remaining alternatives, however. For example, the values of the three central choices are as follows:

Choice	Impact A	Impact B
5	6.0	2.0
6	4.0	4.0
7	3.0	5.0

If we assume the concerns to be of equivalent importance, the impacts for all three choices add to 8.0, and there are no grounds for selecting one over the others. In practice, however, there are often limitations on impacts—soils are buffered against acid rain up to a certain "critical loading," for example, and degrade when that loading is exceeded. If constraints on impacts A and B are imposed on some such basis (Fig. 1.7b), choices 1, 4, and 7 are no longer viable.

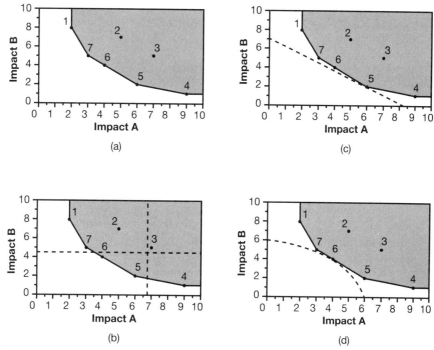

FIGURE 1.7 (a) The impacts of seven possible activity choices on environmental concerns A and B. The gray region is the "decision space" within which choices are possible; (b) the limitation of decision space by the imposition of impact constraints; (c) the interaction of a linear value function with the decision space; (d) the interaction of a nonlinear value function with the decision space. (Adapted with permission from F.C. Field III, J.A. Isaacs, and J.P. Clark, Life cycle analysis and its role in product and process development, International Journal of Environmentally Conscious Design and Manufacturing, 2, 13-20. Copyright 1993 by ECM Press.)

The imposition of impact limitations can produce a single preferred choice, several permissible choices, or no permissible choice at all, partly because the limitations still relate to individual addressing of the impacts. An alternative is to define *value functions* that represent combinations of impacts acceptable to the group evaluating the choices. Two alternative value functions are illustrated in Figures 1.7c and 1.7d; the preferred choice is the point of tangency between the decision space defined by the possible choices and the value functions defined by the permissible impacts.

In this example of multi-objective decision making, the preferred choice (#5 in Fig. 1.7c, #6 in Fig. 1.7d) is clear. The real world and its environment are, of course, more complex. There are many more than two environmental concerns (Table 1.3 lists 17, for example), and the specification of a multi-impact value function for the environment has not yet been approached. Figure 1.7 thus is taking a two-dimensional slice out of a multi-dimensional pie. Nonetheless, the ultimate, if unstated, goal of all life-cycle assessments is to reliably locate the point of tangency, or at least the point of closest approach, between a multi-dimensional value function and a multi-dimensional decision space. If the realization of this goal is difficult (and is, in a sense, the subject of this entire book), the concept remains straightforward.

1.8 APPROACHES TO ASSESSMENT

Engineers, scientists, executives, and administrators make millions of decisions every day that influence the way in which our technological society interacts with the environment. Should the issue under consideration be sufficiently important and the needed expertise, personnel, and financial resources available, a comprehensive LCA has the potential to optimize an industry-environment interaction or a sector-industry environment interaction to the benefit of both. Exploratory studies of these interactions are stimulating a number of commendable actions.

The focus of this book is to describe fully quantitative LCAs, as well as simpler tools, the latter to be used not as guidance for a few decisions but as guidance for millions of them. These subsidiary tools, termed streamlined life-cycle assessments (SLCAs), are admittedly less quantitative and comprehensive than traditional LCAs, but their recommendations to product and process designers are often very similar as well as more timely and less expensive. From the standpoint of environmental burdens, SLCAs have the potential to dramatically improve industry-environment interactions in a way that LCAs do not.

Both LCAs and SLCAs thus have roles to play, and those roles are in the process of becoming better defined. For perspective, this book begins by describing traditional LCA methodology, illustrating its utility, and describing its limitations and deficiencies. SLCAs are then introduced and shown to capture much of the perceived value of LCAs with perhaps minimal disbenefits, as seen by application to a number of industrial products and systems. As shown in some of the later chapters, SLCAs can be applied in a variety of circumstances and to a number of different problems, they can reflect both positive and negative environmental impacts, and they can be used by individuals or assessment teams. LCA/SLCA tools are one part of the approach of our technological society toward sustainability; indeed, they are necessary elements in that journey.

FURTHER READING

Beauchamp, T.L., and N.E. Bowie, *Ethical Theory and Business*, 4th ed. (Englewood Cliffs, NJ: Prentice Hall, 1993).

Beltrani, G., Safeguard subjects: The conflict between operationalization and ethical justification, *International Journal of Life Cycle Assessment, 2* (1997), 45-51.

Berry, R.J., *Environmental Dilemmas: Ethics and Decisions* (London, U.K.: Chapman and Hall, 1992).

Clarke, R., *Water: The International Crisis* (Cambridge, MA: The MIT Press, 1993).

Environmental Protection Agency Science Advisory Board, *Reducing Risk: Setting Priorities and Strategies for Environmental Protection*, Report SAB-EC-90-021 and 021A (Washington, D.C., 1990).

Field, F.R. III, J.A. Isaacs, and J.P. Clark, Life cycle analysis and its role in product and process development, *International Journal of Environmentally Conscious Design and Manufacturing, 2* (1993) 13-20.

Pearce, D.W., *Economic Values and the Natural World* (Cambridge, MA: The MIT Press, 1993).

World Commission on Environment and Development, *Our Common Future* (Oxford, U.K.: Oxford University Press, 1987).

World Energy Council and International Institute for Applied Systems Analysis, *Global Energy Perspectives to 2050 and Beyond* (London: World Energy Council, 1995).

EXERCISES

1.1. The purchase and use of an automobile involves a number of potentially targetable activities, including the use of metals, plastics, and other resources, the consumption of fossil fuels, and the emission of CO_2, NO_x, and lead. Identify the concerns for which these activities apply, and the grand objectives with which the concerns are associated.

1.2. Survey 10 to 20 friends and acquaintances and generate your own list of grand objectives. Is it different from that given in this chapter? Compare and comment on the two lists.

1.3. The "Natural Step" conditions are used as general guidance by a number of corporations, yet they have sometimes been criticized as overly simplistic. Write a one-page essay on the suitability of the conditions.

1.4. Expand the two-dimensional decision-making diagram of Figure 1.7 into three dimensions, so that the allowable shaded decision area becomes a decision volume. If an industrial activity has three potential environmental impacts, can your diagram (perhaps aided by computer graphics) be useful in arriving at the optimum design and manufacturing approach?

CHAPTER 2
The Concept of Life-Cycle Assessment

2.1 THE LIFE CYCLE OF INDUSTRIAL PRODUCTS

The essence of life-cycle assessment is the examination, identification, and evaluation of the relevant environmental implications of a material, process, product, or system across its life span from creation to waste or, preferably, to re-creation in the same or another useful form. The Society of Environmental Toxicology and Chemistry defines the LCA process as follows:

> The life-cycle assessment is an objective process to evaluate the environmental burdens associated with a product, process, or activity by identifying and quantifying energy and material usage and environmental releases, to assess the impact of those energy and material uses and releases on the environment, and to evaluate and implement opportunities to effect environmental improvements. The assessment includes the entire life cycle of the product, process or activity, encompassing extracting and processing raw materials; manufacturing, transportation, and distribution; use/re-use/maintenance; recycling; and final disposal.

Any analysis of the type described above must explicitly treat all life-cycle stages in a typical complex manufactured product, as shown schematically in Figure 2.1.

- Stage 1, premanufacturing, is performed by suppliers drawing on (generally) virgin resources and producing materials and components.
- Stage 2, the manufacturing operation, and
- Stage 3, product delivery. This stage and the previous one are directly under corporate control.
- Stage 4, the customer use stage, is not directly controlled by the manufacturer but is strongly influenced by how products are designed and by the degree of continuing manufacturer interaction.
- In Stage 5, a product no longer satisfactory because of obsolescence, component degradation, or changed business or personal circumstances is refurbished, recycled, or discarded.

The life-stage outline above assumes that a corporation is manufacturing a final product for shipment and sale directly to a customer. Often, however, a corporation's products are intermediates—process chemicals, steel screws, brake systems—made for sale to and incorporation in the products of another firm. How does DFE apply in these circumstances?

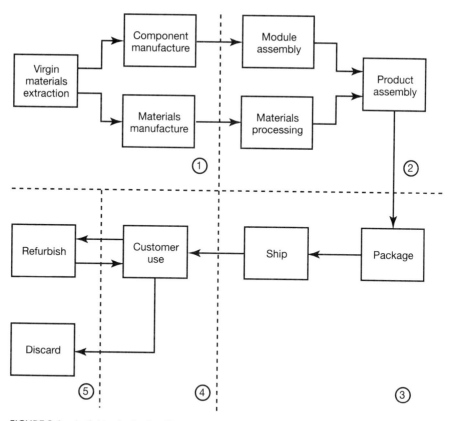

FIGURE 2.1 Activities in the five life-cycle stages (circled numbers) of a product manufactured for customer use. In an environmentally-responsible product, the environmental impacts at each stage are minimized, not just those in Stage 2.

Picture the process of manufacture as shown in Figure 2.2, an expansion of Figure 2.1. Three different types of manufacture are illustrated: Type A is the production of intermediate materials from raw materials (examples: plastic pellets from petroleum feedstock or rolls of paper from bales of recycled mixed paper); Type B is the production of components from intermediate materials (examples: snap fasteners from steel stock or colored fabric from cotton); Type C is the processing of intermediate materials (example: cotton fabric) or the assembly of processed materials (example: plastic housings) into final products (examples: shirts or tape recorders).

Figure 2.1 is for an operation of Type C, where the design and manufacturing team had virtually total control over all product life stages except Stage 1: Premanufacture. For a corporation of Types A or B, the perspective changes for some life stages, but not for others:

- Stage 1, premanufacture. Unless a Type A corporation is the actual materials extractor, the concept of this life stage is identical for corporations of Types A, B, and C.
- Stage 2, manufacture. The concept of this life stage is identical for corporations of Types A, B, and C.

20 Chapter 2 The Concept of Life-Cycle Assessment

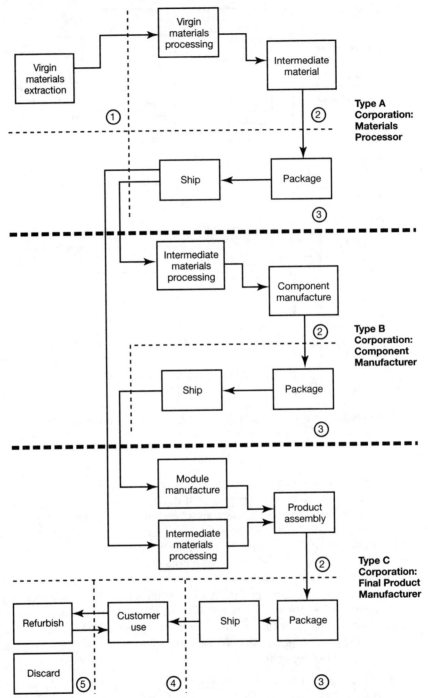

FIGURE 2.2 The interrelationships of product life stages for corporations of Type A (materials processors), Type B (component manufacturers), and Type C (final product manufacturers).

- Stage 3, product delivery. The concept of this life stage is identical for corporations of Types A, B, and C.
- Stage 4, product use. For Type A corporations, product use is essentially controlled by the Type B or C receiving corporation, although factors such as intermediate materials purity or composition can influence such factors as by-product manufacture and residue generation. For Type B corporations, their products can sometimes have direct influence on the in-use stage of the Type C corporation final product, as with energy use by cooling fans or lubricant requirements for bearings.
- Stage 5, refurbishment, recyling, or disposal. The properties of intermediate materials manufactured by Type A corporations can often determine the potential for recyclability of the final product. For example, a number of plastics are now formulated with the goal of optimizing recyclability. For Type B corporations, the approach to the fifth life stage depends on the complexity of the component being manufactured. If it can be termed a part, such as a capacitor, the quantity and diversity of its materials and its structural complexity deserve review. If it can be termed a component or module, the concerns are the same as those for a manufacturer of a final product—ease of disassembly, potential for refurbishment, and the like.

Thus, Type A and B corporations can and should deal with DFE assessments of their products much as should Type C corporations. The considerations of the first three life stages are, in principle, completely under their control. For the last two life stages, the products of Type A and B corporations are influenced by the Type C corporation with which they deal and, in turn, their products influence the life stage 4 and 5 characteristics of Type C products.

2.2 THE LCA FRAMEWORK

A life-cycle assessment is a large and complex effort, and there are many variations. Nonetheless, there is general agreement on the formal structure of LCA, which contains three stages: *goal and scope definition, inventory analysis*, and *impact analysis*, each stage being followed by *interpretation of results*. The concept of the life-cycle methodology is pictured in Figure 2.3. First, the goal and scope of the LCA are defined. An inventory analysis and an impact analysis are then performed. The interpretation of results at each stage guides an analysis of potential improvements, which may feed back to influence any of the stages, so that the entire process is iterative. Finally, the "design for environment" guidance to the technological activity is released.

There is perhaps no more critical step in beginning an LCA evaluation than to define as precisely as possible the evaluation's goal and scope: what materials, processes, or products are to be considered, what are the characteristics and limitations of the study, and how broadly will alternatives be defined. Consider, for example, the question of releases of chlorinated solvents during a typical dry-cleaning process. The goal of the analysis is to reduce environmental impacts from the process. The scope of

Life-cycle assessment framework

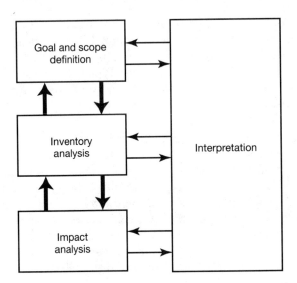

FIGURE 2.3 Stages in the life-cycle assessment of a technological activity. The wide arrows indicate the basic flow of information. At each stage, results are interpreted, thus providing the possibility of revising the environmental attributes of the activity being assessed. (Adapted from Society of Environmental Toxicology and Chemistry (SETAC), Guidelines for Life-Cycle Assessment: A Code of Practice, Pensacola, FL, 1993.)

the analysis, however, must be defined clearly. If it is limited, the scope would comprehend end-of-pipe controls, administrative controls, and process changes. Alternative materials—in this case, solvents—might be considered as well. If, however, the scope is defined broadly, it could include alternative service options: some data indicate that a substantial number of items are sent to dry-cleaning establishments not for cleaning per se but simply for pressing. Accordingly, offering an independent pressing service might reduce emissions considerably. One could also take a systems view of the problem: Given what we know about polymers and fibers, why are clothing materials and designs still being provided that require the use of chlorinated solvents for cleaning? Among the considerations that would influence the choice of scope in cases such as the above are: (1) who is performing the analysis, and how much control they can exercise over the implementation of options; (2) what resources are available to conduct the study; (3) what information will be readily available, and (4) what is the most limited scope of analysis that still provides for adequate consideration of the systems aspects of the problem.

The resources that can be applied to the analysis should also be scoped. Many traditional LCA methodologies provide the potential for essentially open-ended data collection—and, therefore, virtually unlimited expenditure of resources. As a general rule, the depth of analysis should be keyed to the degrees of freedom available to make meaningful choices among options and to the importance of the environmental or technological issues leading to the evaluation. For example, an analysis of using different plastics in the body of a currently marketed portable disk player would probably not require a complex analysis; the degrees of freedom available to a designer in such a case are already quite limited because of the constraints imposed by the existing design and its market niche. On the other hand, a government regulatory organization contemplating limitations on a material used in large amounts in numerous and diverse manufacturing applications would want to conduct a fairly comprehensive analysis, because the degrees of freedom involved in finding substitutes could be quite numerous and the environmental impacts of substitutes implemented widely throughout the economy could be significant.

The second component of LCA, inventory analysis, is by far the best developed. It uses quantitative data to establish the levels and types of energy and materials input to an industrial system and the product output and environmental releases that result, as shown schematically in Figure 2.4. The approach is based on the idea of a family of materials budgets, in which the analyst measures the inputs and outputs of energy and resources as well as the resources embedded in the resulting products and by-products, both those resource flows with value and those that are potential liabilities. The assessment is ideally done over the entire life cycle—materials extraction, manufacture, distribution, use, and disposal.

The third stage in LCA, the impact analysis, involves relating the outputs of the system to the impacts on the external world into which those outputs flow, or, at least, to the burdens being placed on the external world. Aspects of this difficult and often contentious topic are discussed later in the book.

The interpretation of results phase is where the findings from one or more of the three stages are used to draw conclusions and recommendations. The output from this

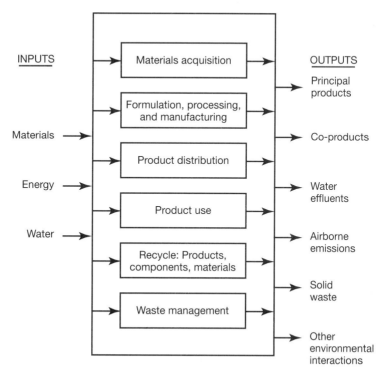

FIGURE 2.4 The elements of a life cycle inventory analysis. (Adapted from Society of Environmental Toxicology and Chemistry (SETAC), A Technical Framework for Life-Cycle Assessment, Washington, D. C., 1991.)

24 Chapter 2 The Concept of Life-Cycle Assessment

activity is often the explication of needs and opportunities for reducing environmental impacts as a result of industrial activities being performed or contemplated. It follows ideally from the completion of stages one through three, and occurs in two forms: a proactive one termed *Design for Environment* (*DFE*), and a reactive one (i.e., one that addresses regulatory constraints) termed *pollution control*. Less comprehensive but still valuable actions can also be stimulated by the results of the scoping and inventory stages.

2.3 THE INDUSTRIAL ECOLOGY FLOW CYCLE

In understanding where a material goes after it enters the materials flow system, it is useful to look at its total budget from the time it is extracted from the ground or harvested until the time it returns to the reservoir of its origins. The first steps in that cycle, the industrial production of the materials themselves, are shown in Figure 2.5.

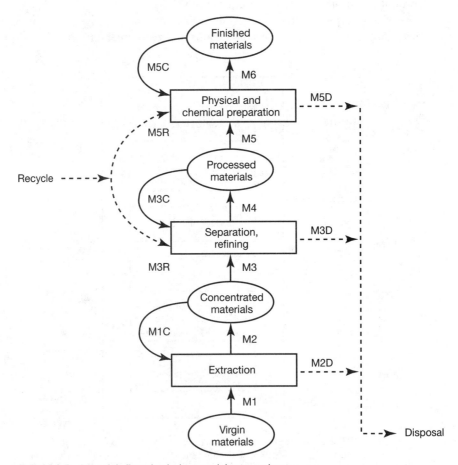

FIGURE 2.5 Materials flows in virgin materials processing.

Beginning with virgin materials, the flows proceed through cycles of extraction, separation and/or refining, and physical and chemical preparation to produce finished materials. A typical example might be the extraction of copper ore from the ground and the eventual production of copper wire.

In addition to the central flows (Mx), Figure 2.5 indicates the materials flows that occur away from the central spine. To the right are the wastes consigned to disposal (MxD), typically large fractions in the early stages of the flow cycle and smaller fractions at later stages. To the left are flows of recycled material (MxR), such as copper wire recovered from obsolete power distribution systems. Near the center are recycled flows that occur during the production process itself (MxC), as when scrap from one process stage is reused in the preceding stage. A complementary flow occurs for any chemicals used in processing the material.

The flows of materials and processes as shown in Figure 2.5 occur within what has traditionally been termed the "heavy industry" community, and the flows are generally under the sole control of the "materials supplier."

The resource flows involved in the manufacture of products from the finished materials produced by the materials supplier are shown in Figure 2.6. As in Figure 2.5, waste flows, recycled material flows, and in-process recycle flows are indicated. An important distinction between this figure and Figure 2.5 is that several finished materials are generally involved, rather than a single one. A typical example is the production of a cable connector for a computer from selected metal and plastic.

It is instructive to examine Figure 2.6 from the perspective of constraints to optimizing the environmental aspects of the process. At the forming step, material may

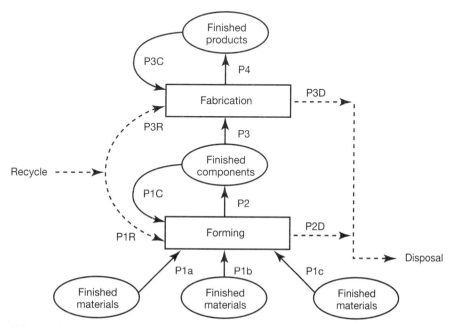

FIGURE 2.6 Materials flows in manufacturing.

enter from three types of streams: the virgin materials streams (P1x), the process recycle streams (P1C), and the recycled material streams (P1R). Unrecycled waste exits in disposal stream P2D. Optimization involves decreasing or eliminating P2D and utilizing whatever external recycled material is available (P1R) within the constraints of product performance requirements, customer preferences, and existing cost structures. The use of any recycled material stream thus involves a tradeoff between ensuring its purity and suitability against the cost of using virgin finished materials. A similar analysis applies to each step of the process.

The flows of materials and processes shown in Figure 2.6 are within the "manufacturing" community, and the materials flows are generally under the sole control of the "manufacturer."

The resource flows involved in the customer portion of the materials cycle are shown in Figure 2.7. Optimization opportunities here include the need to avoid resource dissipation, especially in the handling of obsolete products. To the extent that customers favor the recycle stream C3R at the expense of the waste disposal stream C3D, materials retention is improved. Although exceptions exist, the situation under existing economic and governmental constraints is that customer decisions are made independently of either the material supplier (who may also be a material recycler) or the manufacturer.

When Figures 2.5–2.7 are combined, they result in the ensemble materials flow cycle shown in Figure 2.8. One might view this figure as an inverted potential diagram, in which the energy expended to achieve a given flow decreases as one moves upward in the diagram. That is, it takes much less energy to recycle materials from one of the higher stages to an intermediate stage than to begin at a lower stage.

Robert Ayres of the European Institute of Business Administration puts the point differently; that the work done on materials at the expense of energy represents society's battle against thermodynamics, and the energy invested per unit of material decreases as one approaches the top of the materials flow chain. This increase in order and usability of processed materials is their embedded utility. A goal of environmentally responsible engineering is the preservation of embedded utility in the materials used by industrial processes. One way in which this can be accomplished is to design and build products that will have long service lives. A second way is to retain in use a large portion of an obsolescent product, recycling and replacing only outdated subsystems and components.

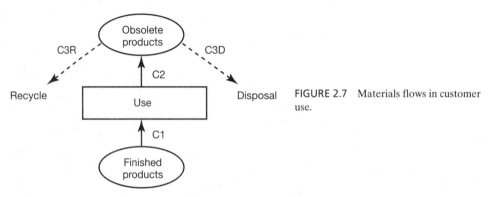

FIGURE 2.7 Materials flows in customer use.

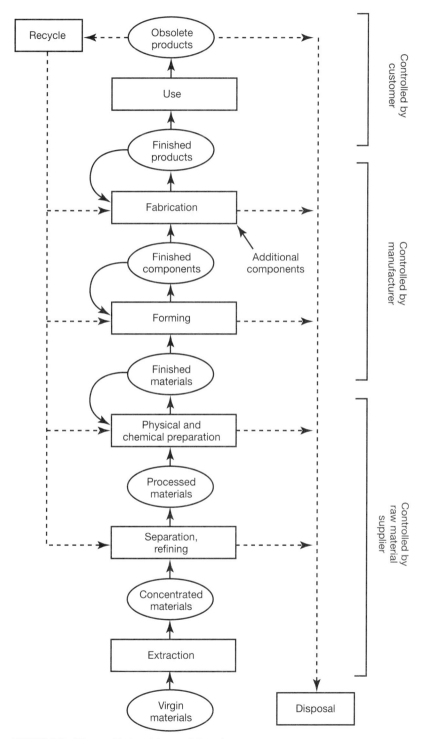

FIGURE 2.8 The total industrial materials cycle.

It is interesting to compare Figures 2.4 and 2.8. The former, dealing with life-cycle assessment, is constructed largely from the perspective of the environmental engineer, whose primary concern is with residues lost to the environment. The latter is constructed largely from the perspective of the manufacturing engineer, whose primary concern is with the flow of materials ending up in products. The responsible technologist, often termed an "industrial ecologist," wants to simultaneously optimize the two concerns, promoting useful technological activity while doing so in ways that cause minimal environmental impact. It is in this spirit that LCA/SLCA is optimally performed.

FURTHER READING

Consoli, F. et al., *Guidelines for Life Cycle Assessment: A Code of Practice*, (Pensacola, FL: Society of Environmental Toxicology and Chemistry (SETAC), 1993).

Keoleian, G.A., and D. Menerey, *Life Cycle Design Guidance Manual: Environmental Requirements and the Product System*, EPA/600/R-92/226 (Washington, D.C.: U.S. Environmental Protection Agency, 1993).

Lindfors, L.-G. et al., *Nordic Guidelines for Life Cycle Assessment* (Copenhagen, Denmark: Nordic Council of Ministers, Report Nord 20, 1995).

Society of Environmental Toxicology and Chemistry (SETAC), *A Technical Framework for Life-Cycle Assessment* (Washington, D. C., 1991).

EXERCISES

2.1. Choose a common but sophisticated household appliance such as a refrigerator, a television set, or a washing machine. Using the diagram in Figure 2.1 and any other information available to you, describe the life stages of the appliance, including who is primarily responsible for the environmental concerns at each life stage.

2.2. You are the LCA analyst for a papermaking company and are asked to do a life-cycle assessment for a new type of paper to be used for printing currency. Define and describe the scope of your assessment.

2.3. Repeat Exercise 2.2 for the situation in which you work for a forest products company that supplies wood fiber for the paper.

2.4. Using Figure 2.5, devise an equation for the percent of discarded materials relative to the amount entering the processing stream.

CHAPTER 3

Goal Setting, Scope Determination, and Inventory Analysis

3.1 GOAL SETTING AND SCOPE DETERMINATION

The most common LCA goal is to conduct an assessment of the environmental attributes of a specific product or process and to derive information from that assessment on how to improve environmental performance. If this exercise is conducted early in the design phase, the goal may be to compare two or three alternative designs. If the design is finalized, the product is in manufacture, or the process is in operation, the goal can probably be no more than to achieve modest changes in environmental attributes at minimal cost and minimal disruption of an existing operation.

It is possible for an assessment target to be much more ambitious than the evaluation of a single product or process. This usually occurs with the evaluation of a system of some sort, the operation of an entire facility or corporation, for example, or of an entire governmental entity. In such a case, it is likely that alternative operational approaches can be studied, but not alternative systems. In addition, a system that makes a logical entity from an LCA viewpoint may involve more than one implementer, so collaborative goal setting may be required. If a goal can be quantified, such as "achieve a 20 percent decrease in overall environmental impact," it is likely to be more useful and the result more easily evaluated. Quantification of the goal requires quantification of each assessment step, however, and quantitative goals should be adopted only when one is certain that adequate data and assessment tools are available.

The scope of the assessment is perhaps best established by asking a number of questions: "Why is the study being conducted?" "How will the results be used, and who will use them?" "Do specific environmental issues need to be addressed?" "What level of detail will be needed?" It is useful to recognize that life-cycle assessment is an iterative process, and that the scope may need to be revisited as the LCA proceeds.

The scope plays a major role in the effort, time, and expense that will be involved in an LCA. If confined to a subset of life stages or a limited selection of impacts, the LCA will be much more efficiently carried out. The result may be a report of limited usefulness, however, if potentially important factors have been omitted. Nonetheless, experienced assessors often opt for restricting the LCA scope in order to increase tractability. Their concept is that significant benefit can often come from a limited LCA. It is, in fact, not assured that more substantial benefits will emerge from a more extensive LCA, and an LCA so broadly scoped that its successful completion is uncertain may turn out to yield no benefits at all.

3.2 DEFINING STUDY BOUNDARIES

The potential complexity of comprehensive LCAs is nowhere better illustrated than by the problem of defining the boundaries of the study. There are many potential issues for discussion in this regard, no national or international standards, and no consensus on which are the best ways to approach boundary issues. The present discussion proceeds by exploring a number of these issues, and concludes with some general recommendations concerning choices of boundaries in LCA.

3.2.1 Life Stage Boundaries

Early attempts to evaluate industrially related environmental impacts focused exclusively on activities within manufacturing facilities themselves. From the life-cycle perspective shown in Figure 2.1, these approaches can be regarded as being restricted to life stage 2, or what today is sometimes termed *gate to gate analysis*, i.e., from the factory gate through which materials enter to the gate where products leave. The present and commendable "pollution prevention" activities have similar limitations.

The environmental performance of products during their use became an issue in the late 1980s. Emissions from automobile exhaust began to be widely controlled, and regulations or guidelines on energy use by appliance and office machines began to be issued. Manufactures thus were encouraged to think about the environmentally related aspects of life stage 4.

In the early 1990s, Germany implemented regulations requiring manufacturers to take back the packaging used for their products: boxes, cushioning foam, plastic wrap, and so on. This encouraged manufacturers to minimize packaging and make it more recyclable, in effect adding life stage 3 to corporate environmental assessment.

Life stage 5 has also begun to come under consideration. Several European countries are considering "takeback" laws or agreements that will compel manufacturers to recover their products when the customer no longer wants them. Some manufacturers are discovering that such recovery, followed by refurbishment and reuse, can be profitable. These actions in turn encourage design and materials selection decisions that optimize the value of recovered products. The result is the incorporation of life stage 5 in environmental planning.

Life stage 1, premanufacture, has proven difficult to address although some attempts have been made. That life stage is outside the direct control of the manufacturer, and acquiring the necessary information often depends on the willingness to share sensitive information. Nonetheless, some of the more environmentally aggressive firms are working with their suppliers to include life stage 1 in their analyses.

3.2.2 Level of Detail Boundaries

How much detail should be included in an LCA? An assessor frequently needs to decide whether effort should be expended to characterize the environmental impacts of trace constituents such as minor additives in a plastic formulation or small brass components in a large steel assembly. With some modern technological products containing hundreds of materials and thousands of parts, this is far from a trivial decision. One way it is sometimes approached is by the *5 percent rule*: If a material or component comprises less than 5 percent by weight of the product, it is neglected in the LCA. A common amendment to this rule is to include any component with particularly severe environmental inputs. For example, the lead-acid battery in an automobile weighs less than 5 percent of the vehicle, but the toxicity of lead makes the battery's inclusion reasonable. Such items (the list could be long) often include mercury relays, chrome-plated parts, and radioactive materials.

3.2.3 The Natural Ecosystem Boundary

In a number of industrial processes, the functions of nature interact with those of the technological society. Consider the process of generating electric power by burning wood, as shown in Figure 3.1. The industrial components of the process are the harvesting of the wood (using fuel oils for cutting and transport) and the combustion process itself. The natural components are the formation of the wood biomass (i.e., the growing of trees) and the biodegradation of the harvesting wastes. Some LCAs would choose to

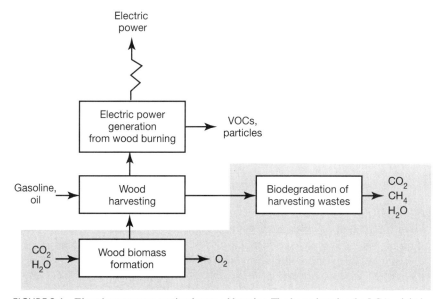

FIGURE 3.1 Electric power generation by wood burning. The boundary for the LCA might be chosen to include the shaded areas or to exclude them. The implications of this choice are discussed in the text. (Adapted from L.-G. Lindfors, et al., Technical Report No. 4, Tema Nord 1995.502. Copyright 1994 by the Nordic Council of Ministers.)

draw the assessment boundary around only the industrial components of the process, while others would include the natural components as well. The latter is more complete, but the inclusion of the natural components is likely to make quantification considerably more difficult.

A second natural ecosystem issue that arises when choosing LCA boundaries is that of biological degradation. When industrial materials are discarded, as into a landfill, biodegradation produces such outflows as methane from paper, chlorofluorocarbons from blown foam, and mobilized copper, iron, and zinc from bulk metals. Approaches to these complications have included incorporating these flows in the LCA inventory, excluding landfill outflows completely, or including those flows for a specified time period. Flows from landfills are generally difficult to estimate, so one is faced with a tradeoff between accuracy and comprehensiveness.

Another example of the natural/industrial boundary issue is the process of making paper from wood biomass, as shown in Figure 3.2. Here the assessor has several levels of possible inventory detail open. The basic analysis is essentially a restriction of the inventory to life stage 2 (refer to Fig. 2.1). The energy envelope incorporates some of the external flows related to the production of energy. The extended envelope includes all life-cycle stages and flows directly connected with the industrial system. The comprehensive envelope adds the natural processes of biomass formation and the degradation of materials in a landfill. None of these options is inherently correct or incorrect, but the choice could lead to quite different LCA results.

3.2.4 Boundaries in Space and Time

A characteristic of environmental impacts is that their effects can occur over a very wide range of spatial and temporal scales. The emission of large soot particles affects a local area, those of oxides of nitrogen generate acid rain over hundreds of kilometers, and those of chlorofluorocarbons influence the planetary ozone layer. Similarly, emissions causing photochemical smog have a temporal influence of only a day or two, the disruption of an ecosystem several decades, and the stimulation of global climate change several centuries. LCA boundaries may be placed at short times and small distances, long times and planetary distances, or somewhere in between. The choice of any of these boundary options in space and time may be appropriate depending on the scope of the LCA.

3.2.5 Choosing Boundaries

It should be apparent that the choice of LCA boundaries can have enormous influence on the time scale, cost, and tractability of the LCA. The best guidance that can be given is that the boundaries should be consistent with the goals of the exercise. An LCA for a portable radio would be unlikely to have goals that encompass impacts related to energy extraction, for example, both because the product is not large and because its energy impacts will doubtless be modest. A national study focusing on flows of a particular raw material might have a much more comprehensive goal, however, and boundaries would be drawn more broadly. The goals of the LCA thus define much of the LCA scope, as well as the depth of the inventory and impact analysis activities.

Section 3.2 Defining Study Boundaries 33

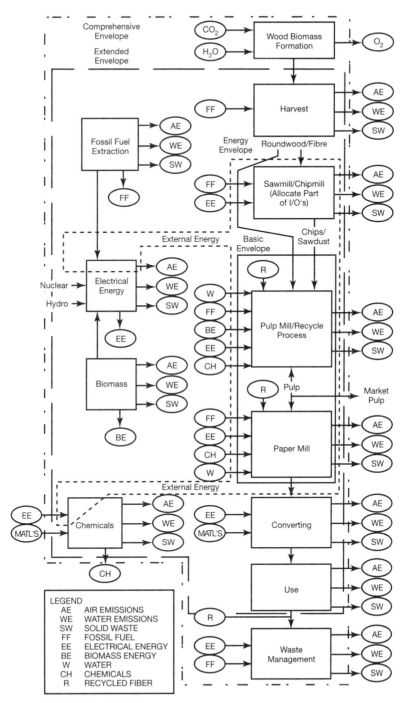

FIGURE 3.2 A simplified quantitative inventory flow diagram for the manufacture of paper. Four levels of possible detail are shown. (Adapted from a diagram provided by courtesy of Martin Hocking, University of British Columbia.)

3.3 THE FUNCTIONAL UNIT

A seemingly arcane topic that can become crucial when actually implemented in LCA is that of the *functional unit*: the amount of product, material, or service to which the LCA is applied. This definition becomes especially important when two rather dissimilar items are being compared. Take, for example, an LCA in which cloth diapers are compared to disposable diapers. The functional unit is presumably one of each type of diaper. Complications ensue, however. The first is that cloth diapers can be washed either commercially or at home, with very different energy requirements, so each of these options should be included in the study. More important is that cloth diapers are often used two or three at a time (average use is 68 per baby-week) while disposables are more leak-proof and never used in multiples (average use is 38 per baby-week). The correct functional unit is thus one *diapering*, not one diaper.

Another example is provided by an LCA of hot-drink cups, which may be coated paper, blown polystyrene, or ceramic. One's first intuition is to compare one of each kind of cup. However, the first two types are discarded after a single use, while ceramic cups may be used hundreds of times. The more logical unit is thus *a single use* of a hot-drink cup. This approach gains still more credence when one considers that paper and polystyrene cups are often used two at a time to prevent users from burning their hands. Thus, a single use requires something between one and two cups on average.

A final illustration of the subtlety sometimes involved in the selection of a functional unit is an LCA of refrigerators conducted in Sweden in 1994, in which a standard refrigerator was compared to one designed to have twice the insulating capability and to be longer-lived. The instinctive definition of the appropriate functional unit is that it is one refrigerator. However, that choice overlooks the facts that refrigerators must have similar exterior dimensions (to fit standard spaces in housing units), and thus the better-insulated refrigerator has less cooled volume. Providing a certain amount of cooled volume is, of course, the function that is desired, so one cubic meter of cooled volume is a better functional unit than one refrigerator. Still neglected, however, is the anticipated longer life of the redesigned unit. To capture that difference, the preferable functional unit was thus suggested to be one cubic meter-year of cooled volume.

The choice of the functional unit is therefore not straightforward, and it requires a knowledge of the way in which the product, material, or service will be used, the actual function that is being satisfied, and perhaps even the anticipated longevity of the item being assessed. A good choice can support a useful LCA outcome; a poor choice may render the LCA essentially meaningless.

3.4 APPROACHES TO DATA ACQUISITION

After the scale of the LCA has been established by defining the scope, the analyst proceeds to the acquisition of the necessary data. The process is begun by constructing, in cooperation with the design and manufacturing team, an inventory flow diagram. The aim is to list, at least qualitatively but preferably quantitatively, all inputs and outputs of materials and energy throughout most or all life stages. Figure 3.3 shows a typical example of such a diagram for the manufacture of a portable radio in which the hous-

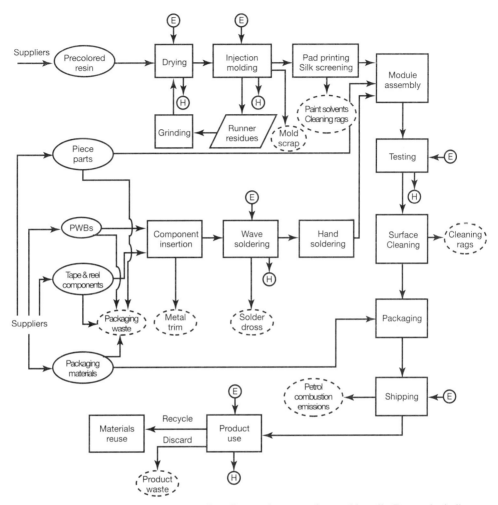

FIGURE 3.3 A qualitative DFE inventory flow diagram for a generic portable radio. Rectangles indicate process flow steps; ellipses, material or subassembly flows into the process; dashed ellipses, material flows to waste disposal; parallelograms, material flows to recycling processes. Significant energy use is indicated by a circled E, and significant heat loss by a circled H.

ing is molded in the plant from precolored resin, the electronics boards are constructed from components furnished by suppliers, and those pieces and others (speaker, electronic jacks, batteries, etc.) are assembled into the final product. The diagram depicts a practice often followed for reasons of practicality and efficiency at a sacrifice of completeness: the flows of materials and energy are indicated for the final four life stages, but not the premanufacture (supplier) stage. The diagram indicates a number of material and energy byproducts (the latter being mostly unused heat). After the inventory flow diagram is constructed—in as much detail as possible—the actual inventory analysis can begin.

Some of the information needed for an inventory analysis is straightforward, such as the amounts of specific materials needed for a given design or the amount of cooling water needed by a particular manufacturing process. Quantitative data obviously have advantages: they are universally utilized in high technology cultures, they offer powerful means of manipulating and ordering data, and they simplify choosing among options. However, the state of information in the environmental sciences may not permit the secure quantification of environmental and social impacts because of fundamental data and methodological deficiencies. In some instances, those concerns that resist quantification are simply ignored—thereby undercutting the systemic approach inherent in the design for environment concept.

In some cases, the information that is available may not be of the traditional quantitative type, but may still prove very useful. Accordingly, the analyst should be willing to approach data needs with a broad perceptive. Qualitative information, whether it applies to materials selection, processes, components, or complex products, can often be as effective as quantitative data. A qualitative approach can be somewhat controversial among engineers and business planners, both of whom are biased toward quantitative systems. The utility of qualitative systems is great, however, and they are especially valuable when the alternative is to do nothing. Qualitative approaches thus form one of the central themes of this book.

In order to maximize efficiency and innovation and avoid prejudgment of normative issues, an LCA information system should be non-prescriptive. It should provide information that can be used by individual designers and decision makers given the particular constraints and opportunities they face, but should not, at early stages of the analysis, arbitrarily exclude possible design options. For example, use of highly toxic materials might be a legitimate design choice—and an environmentally preferable choice from among the alternatives—when the process designer can adopt appropriate engineering controls. Designing products and processes inherently requires balancing many considerations and constraints, and the necessary tradeoffs can only be made on a case-by-case basis during the product realization process.

LCA information should provide not only relevant data but, if possible, also the degree of uncertainty associated with that data. This approach is particularly important in the environmental area, where uncertainty about risks, potential costs, and potential natural system responses is endemic.

3.5 PROCESS AND PRODUCT BUDGETS

One of the more straightforward materials budgets is that for a manufacturing process. A schematic example is pictured in Figure 3.4, which shows a chemical process involving the cleaning of a product or product component with a liquid solvent. The process begins with the addition of new solvent to a solvent reservoir, followed by piping or otherwise moving the solvent to the product line where the solvent wash occurs. Most of the solvent eventually enters a recycling (and/or disposal) stream, but a portion (known as "drag out") is retained on the product. Some of the drag out material remains on the product, while a fraction is lost to the atmosphere by evaporation.

Mass balance equations can be set up around any boundary of the system. For example, it is clear that the rate at which the solvent leaves the facility must be equal to the rate at which it enters, i.e.,

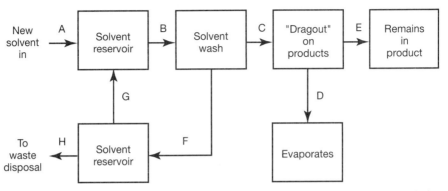

FIGURE 3.4 Schematic diagram of the flow streams involved in a budget analysis of a chemical solvent in a solvent washing process. Capital letters indicate the flows of material.

$$A = D + E + H \tag{3.1}$$

Once this diagram is drawn, the analyst can begin to quantify the budget. The amount of solvent entering the system is probably known, and the rate of solvent purification may be known as well. The rates of loss may or may not be readily quantified. It is clear, however, that if most of the rates are known, some of the others can be computed. If one does not worry too much about absolute accuracy, but is content with rough estimates, an approximate budget can often be put together relatively quickly and simply.

Product budgets share many of the characteristics of process budgets, as shown in the schematic example in Figure 3.5. The first step is to choose a material of interest, generally either one of the predominant materials in the product or one for which significant environmental impact is possible. The next step is to follow that material through the manufacturing process and determine what fraction ends up in the product, what fraction is recycled within the manufacturing facility, and what fraction is lost to disposal. For example, the amount of polymer used in the product manufacturing stream must be equal to that leaving the facility as part of the product plus that leaving as part of the residue stream, or

$$A = D + H_M + H_T + H_I = D + J \tag{3.2}$$

Attempting to construct a budget for such a process may demonstrate, for example, that the flow rate of the total residue disposal stream is known but the individual flows from the manufacturing stages are not. Measurements or estimates can then suggest the most important process stage on which to focus if one wants to reduce the residue disposal stream.

The materials flow diagrams that were given in Figures 2.5–2.7 turn out to be strikingly similar to the multi-stage countercurrent cascades commonly encountered in chemical engineering, and can be analyzed by similar mathematical approaches. The analogy can be appreciated by redrawing Figure 3.5 so that it takes the form shown in Figure 3.6. In this diagram, the product inspection steps are treated as process stages that do not receive countercurrent flow. A materials balance can then be set up for each stage or for the entire system, and overall efficiencies can be computed.

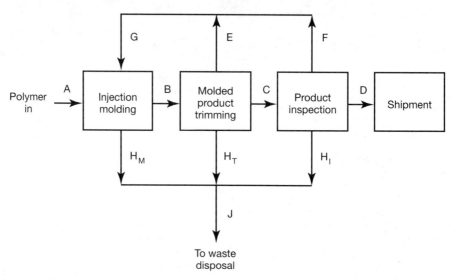

FIGURE 3.5 Schematic diagram of the flow streams involved in a budget analysis of a polymer in an injection molded product.

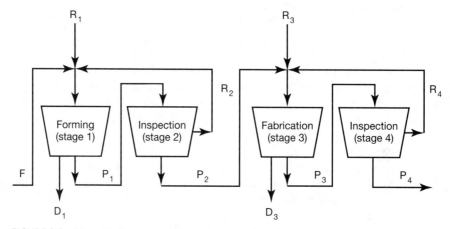

FIGURE 3.6 Materials flows in manufacturing treated as a four-stage countercurrent flow system. All symbols on the diagram refer to total mass flows: F = input flow, P = product flows, D = discard flows, R = recycle flows. See text for additional discussion.

For stage 1, the materials balance is written as

$$F + R_1 + R_2 = P_1 + D_1 \tag{3.3}$$

and the extraction efficiency ε (for stage 1, for example) is given by the ratio of the output to the input:

$$\varepsilon_1 = \frac{P_1}{P_1 + D_1} \tag{3.4}$$

In the same manner, the mass balance equations for stages (2)–(4) are

$$P_1 = P_2 + R_2 \tag{3.5}$$

$$P_2 + R_3 + R_4 = P_3 + D_3 \tag{3.6}$$

$$P_3 = P_4 + R_4 \tag{3.7}$$

The reuse efficiency ρ for an inspection stage is given by (stage 2, for example)

$$\rho_2 = \frac{P_2}{P_2 + R_2} \tag{3.8}$$

and the efficiencies for the entire process are cascaded:

$$\varepsilon = \varepsilon_1 \varepsilon_3 \tag{3.9}$$

$$\rho = \rho_2 \rho_4 \tag{3.10}$$

How the computation of process efficiencies works in practice can be illustrated by an example. Assume for the manufacturing system shown in Figure 3.4 that the residue stream from stage 3 is 5 percent of the product stream from that stage, that the internal recycling streams are 3 percent of the product streams, and that the external recycling streams are 20 percent of the internal process input streams. Then

$$D_3 = 0.05P_3 \quad R_2 = 0.03P_2 \quad R_4 = 0.03P_4$$
$$R_1 = 0.2F \quad R_3 = 0.2P_2$$

Now evaluate the process if the input flow stream to the facility is 100 kg/hr and the output flow is 130 kg/hr. Performing mass balances around the various stages gives (approximately)

$$P_3 = 133.9 \text{ kg/hr} \quad D_1 = 6.1 \text{ kg/hr}$$
$$\varepsilon_1 = 0.95 \quad \varepsilon = 0.91 \quad \rho_2 = 0.97 \quad \rho = 0.94$$

It is straightforward to extend this computation to treat an individual material in streams of mixed materials, or to determine efficiencies over the combined product life cycle rather than just for the manufacturing stage.

3.6 ALLOCATION

Industrial facilities often manufacture more than one product on a single production line using ensemble incoming streams of resources and generating ensemble outputs of residue streams (such a situation is illustrated in Figure 3.7a). The question is how the environmental impacts of the atmospheric, waterborne, and solid residues should be allocated in the LCAs of products A and B. A relatively simple technique is to perform the allocation based on relative product weight, as shown in Figure 3.7b. An alternative technique is to link economic benefits and environmental impacts by allocating on the basis of relative value, not relative weight.

(a)

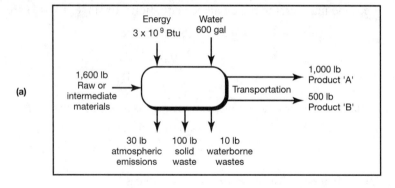

(b)

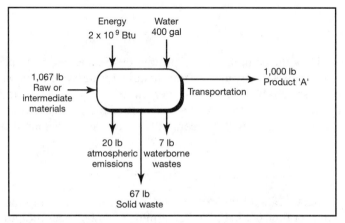

Coproduct Allocation for Product 'A'

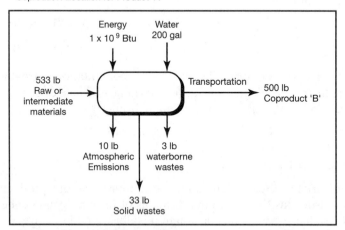

Coproduct Allocation for Product 'B'

FIGURE 3.7 The allocation of inventory flows for a process in which two co-products are manufactured. (a) The actual inventory diagram; (b) the allocation, based on relative product weight. (B.W. Vigon et al., *Life-Cycle Assessment: Inventory Guidelines and Principles*, Rpt. EPA/600/R-92/245 (Cincinnati, OH: U.S. Environmental Protection Agency, 1993).

Allocation can become more intricate in cases like open-loop recycling, in which a product C, after being used, serves as input to a different type of product D: mixed plastic components used to make plastic fence posts, for example. Should an LCA of the plastics components anticipate open-loop recycling and be charged a portion of the impacts of the recycling process, or should product D assume those burdens? Should the burdens be allocated between C and D in some way, and, if so, how? At present there is no really satisfying way of settling these issues, and no consensus standard to use for guidance.

Working groups studying allocation do not resolve these issues, but have made some useful recommendations:

- Whenever possible, allocation should be avoided or minimized, for example by dividing processes into subprocesses or by altering system boundaries.
- Whenever possible, allocation should be based on actual inventory data from the process rather than on the basis of secondary data such as weight or value.
- If several equally reasonable allocation choices exist, a sensitivity analysis should be used to determine the consequences of one choice or another.

3.7 CASE STUDY: DETERGENT-GRADE SURFACTANTS

Surfactants are constituents of detergents that aid in releasing soil from clothing, linens, and other items. Various surfactants can be used, and those surfactants are made from various sources of raw materials. To evaluate whether some options for surfactant sourcing and production were preferable to others, Procter & Gamble Company, in cooperation with Franklin Associates, performed an extensive life-cycle inventory study.

To begin the study, production flow diagrams from raw materials to products were developed. An example, using palm fruit as the principal raw material, is shown in Figure 3.8. Crude oil and natural gas are also needed to synthesize the desired product, a family of alcohol ethoxylates (AE). For each flow of materials, the mass requirements were measured and emissions (not shown on the diagram) were determined. Similar assessments were made for the production of surfactants manufactured from petrochemicals, palm kernel oil, and tallow. A selection of the results is shown in Table 3.1.

The results show more similarities than differences, but the differences are worth comment. One is the much larger water consumption in the case of tallow, attributable to the crop irrigation required to provide feed for beef cattle (beef cattle are the primary source of tallow). The energy requirements for petrochemical recovery are higher than those for palm oil or tallow. Particulates and land-applied emissions from the petrochemical feed stock are significantly lower than for the other options. The conclusion of the study was that benefits from one process appeared offset by liabilities in another, and environmental concerns did not support any fundamental shifts in the worldwide mix of feedstocks used for surfactant manufacture.

The implications of studies such as this are much larger than which feedstock is preferable. Indeed, they have potential application to all industrial operations, both those dealing with components entering the manufacturing operation and with products leaving the manufacturing cycle. In practice, they lend encouragement to pollution

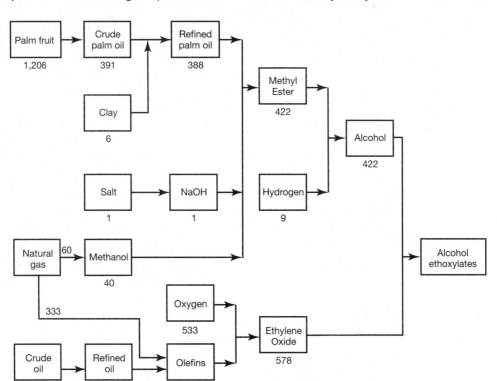

FIGURE 3.8 The flow of materials for the production of detergent surfactants from palm oil. Mass requirements (kg) are expressed on the basis of 1000 kg alcohol ethoxylate produced. (Reproduced with permission from C.A. Pittinger, J.S. Sellers, D.C. Janzen, D.G. Koch, T.M. Rothgeb, and M.L. Hunnicutt, Environmental life-cycle inventory of detergent-grade surfactant sourcing and production, *Journal of the American Oil Chemists' Society, 70*, 1-15. Copyright 1993 by The American Oil Chemists' Society.)

prevention, the effort to reduce or eliminate the generation of pollutants as a consequence of industrial activity. Pollution prevention is a useful goal often encapsulated by the phrase "less is better." However, the feedstock assessment and the spectrum of pollution prevention activities do not face the really crucial issue in implementing LCA in an industrial setting: Assigning relative impact values to each of the materials flow and energy use comparisons made in this analysis. Without such assignments, one is implicitly assuming that the emission of a kilogram of wastewater from a facility is no more and no less important from an environmental standpoint than is the emission of a kilogram of ozone-depleting gas or a kilogram of toxic solid waste. Obviously, all environmental impacts are not equal, and attempting to evaluate them on a relative basis brings regulatory, legal, environmental, corporate, and social factors into play. This crucial LCA step is that of impact analysis, which is the subject of Chapter 4.

FURTHER READING

Fava, J. et al., Eds., *Life-Cycle Assessment Data Quality: A Conceptual Framework* (Pensacola, FL: Society for Environmental Toxicology and Chemistry, 1994).

TABLE 3.1 Raw Material Requirements, Energy Requirements, and Net Life-Cycle Emissions for AE Production[@]

Flow Stream	Feedstock		
	Petrochemical	Palm Oil	Tallow
Organic raw matls.[*]	935	899	891
Water use[*]	40	49	415
Energy use[†]			
Raw matls.	50	26	27
Transport	3	5	7
Processing	39	37	40
Atmospheric emissions[‡]			
Particles	2.2	9.0	8.0
Hydrocarbons	39.2	33.2	34.8
Aqueous emissions[‡]			
Dissolved solids	5.6	5.3	4.3
Oil	0.065	0.12	0.30
Land-applied emissions[‡]	76	111	139

[*]kg per 1000 kg of raw materials.
[†]GJ of energy per 1000 kg of raw materials.
[‡]kg of emissions per 1000 kg of surfactant.
[@]Abstracted from C.A. Pittinger, J.S. Sellers, D.C. Janzen, D.G. Koch, T.M. Rothgeb, and M.L. Hunnicutt, Environmental life-cycle inventory of detergent-grade surfactant sourcing and production, *Journal of the American Oil Chemists' Society, 70*, 1-15, 1993.

Huppes, G., and F. Schneider, Eds., *Proceedings of the European Workings on Allocation in LCA*, ISBN 90-519-078-9 (Leiden: Center of Environmental Science, 1994).

Kim, S., T. Huang, and K.M. Lee, Allocation for cascade recycling system, *International Journal of Life Cycle Assessment, 2* (1997) 217–222.

Klopffer, W., Allocation rule for open-loop recycling in life cycle assessment—A review, *International Journal of Life Cycle Assessment, 1* (1996) 27-31.

Kluppel, H. J., Goal and scope definition and life cycle inventory analysis, *International Journal of Life Cycle Assessment, 2* (1997) 5-8.

Portney, P.R., The price is right: Making use of life cycle analyses, *Issues in Science and Technology, 10* (1993-94) 69-75.

Vigon, B.W., D.A. Tolle, B.W. Cornaby, H.C. Latham, C.L. Harrison, T.L. Boguski, R.G. Hunt, and J.D. Sellers, *Life-Cycle Assessment: Inventory Guidelines and Principles*, EPA/600/R-92/036 (Cincinnati, OH: U.S. Environmental Protection Agency, 1992).

EXERCISES

3.1. Choose one of the following products: a bar of soap, a bicycle, a car wash, an ocean cargo, a hamburger. For each, construct a materials and energy flow diagram of the type shown in Figure 3.3, but with minimal detail in the manufacturing stage and enhanced detail in the product use stage. Suggest appropriate actions by the design engineer for each of those items.

3.2. What are appropriate space and time boundaries for the following:
  (a) A manufacturing process that emits halogenated gases harmful to the stratospheric ozone layer.
  (b) Ecosystem disruption resulting from major facilities construction on the edge of a large river.
  (c) The use of radionuclides in medical radiation therapy.

3.3. Define a suitable functional unit for the following products or services:
  (a) Fluorescent and incandescent light bulbs
  (b) Ferrying cars across a river
  (c) Manufacture and use of an article of clothing

3.4. Repeat the allocation process of Figure 3.7 for the case in which
  (a) The process generates equivalent amounts of Products A and B.
  (b) The monetary value of Product B is five times that of Product A.

CHAPTER 4

Impact and Interpretation Analysis

4.1 LCA IMPACT ANALYSIS

The previous chapter discussed the component of LCA called inventory analysis. Quantitative information on materials and energy flows is acquired at that stage in some cases, qualitative information in other cases. In the data presentations in the previous chapter, it was obvious that some aspects of life-cycle stages have the potential to be more environmentally problematic than others, but the approach begged the question of priorities. One can easily foresee a situation in which alternative designs for a product or process have similar materials use rates but use different materials. How does the analyst make a rational, defendable decision among such alternatives? The answer is that (1) the environmental influences of the activities revealed by the LCA inventory analysis on specific environmental properties must be accurately assessed, and (2) the relative changes in the affected environmental properties must be given some sort of a priority ranking. Together, these steps constitute LCA's *impact analysis.*

Assessing environmental influences is a complicated procedure, but it can, in principle at least, be performed by employing *stressors*, which are items identified in the inventory analysis that have the potential to produce changes in environmental properties. For example, lead in motor vehicle exhaust is a stressor because lead is a systemic poison for animals and humans. The relationships among stressors and the environment are developed by the environmental science community, and are not always available with the degree of detail and precision needed. Conceptually, however, the needed stressor relationships can be imagined to be established and available for use. By combining LCA inventory results with stressor relationships, a manufacturing process might be found, for example, to have a minimal impact on local water quality, a modest impact on regional smog, and a substantial impact on stratospheric ozone depletion.

Thus, the first step in LCA impact assessment is the identification of stressors and the environmental concerns to which they are related. In the simpler of two possible approaches, and the one more often followed, the result of this stage is an estimation of burdens, that is, of *potential impacts.* These potential impacts do not take into account the actual sensitivity of the receptor, and cannot therefore be verified by measurement.

The more challenging approach is to predict *actual impacts*, which, in principle, can be verified by measurement. In either approach, but especially the latter, close collaboration with environmental scientists is required. Moreover, this collaboration must take place across national boundaries, because multinational corporations need to deal with countries that have widely differing priority rankings of environmental properties. A major complicating factor is that risk prioritization is not merely scientific, it also reflects the value system of the community performing the prioritization. Prioritization thus requires harmonization across cultures as well as agreement on scientific and technological issues.

The impact assessment can be structured according to the following steps:

- *Classification.* Classification begins with the raw data from the inventory analysis on flows of materials and energy. Given that data, the classification step consists of identifying environmental concerns suggested by the inventory analysis flows. For example, emissions from an industrial process using a petroleum feedstock may be known to include methane, butene, and formaldehyde. Classification assigns the first primarily to global warming, the second to smog formation, the third to human toxicity (although there may be small impacts on other concerns as well).
- *Characterization.* Characterization is the process of combining different stressor-impact relationships into a common framework. An example is the use of ozone depletion potentials, in which the effects of a molecule of one substance on stratospheric ozone are compared quantitatively with those of another.
- *Localization.* Localization is the operation of comparing environmental impacts occurring in different regions with different characteristics. For example, the process of localization attempts to compare the emission of 1 kg of moderately toxic material into a pristine ecosystem with the impact of the same emission into a highly polluted ecosystem. Two considerations are involved. The first is the relationship of emissions from the product or process being assessed relative to all similar emissions in the region. The second is the degree to which the region possesses assimilative capacity for the emittant. Examples of approaches to this step are provided later in this chapter. (Localization has sometimes been called "normalization," but the word localization is more descriptive. It is also less confusing, because the common mathematical process of normalizing data to different scales is sometimes used in LCA but is unrelated to the activity being described here.)
- *Valuation.* Valuation is the process of assigning weighting factors to the different impact categories based on their perceived relative importance as set by social consensus. For example, an assessor or some international organization might choose to regard ozone depletion impacts as twice as important as loss of visibility, and apply weighting factors to the normalized impacts accordingly.

Although it has been much more common for environmental scientists to study impacts individually rather to attempt to rank them in priority order, several different approaches to prioritization have been made, and the rank-ordering of environmental problems that emerges from these efforts is discussed below. The purpose of the present chapter is not to present or defend specific ordering of impacts, but to illustrate the process.

4.2 LINKING OF STRESSORS AND IMPACTS

An important consideration that plays a role in the prioritization of impacts is that a single impact generally has many stressors, and a single stressor has many impacts. As a consequence, the industry-environment interaction generally embodies a summation of impacts of different magnitudes and different spatial and temporal scales. The result is that a choice between two design options may involve tradeoffs such as choosing between having a modest influence on a single impact of high importance or smaller influences on several impacts of lower importance. A recent example of this complexity involved the substitution in the electronics industry of organic solvents for several CFC cleaning operations; this action constituted the substitution of modest impacts on photochemical smog, increased energy consumption, aqueous residue streams, and air toxics in place of a substantial impact on stratospheric ozone because the latter was judged to be of overriding importance.

One approach to linking stressors and impacts is a series of matrix displays, the axes of the matrices being Technological Activities (specific activities or industries that generate stressors such as gases or particles) and Critical Properties (specific impacts). A matrix of this type for atmospheric concerns is shown in Figure 4.1. To construct this figure, a critical atmospheric concern such as "precipitation acidification" and its direct and indirect chemical causes are linked with the sources responsible for initiating those interactions. The result is a matrix that shows the impact of each potential source of atmospheric change on each critical atmospheric component. A rough qualitative assessment of significance is included. The diagram also includes estimates of reliability, an important component of an assessment effort. The matrix is constructed with separate rows for industrial operations and for the major energy generating processes, and can obviously be expanded in much greater detail than is shown here.

The simplest impact assessments of Figure 4.1 involve only a single cell of the matrix. A typical example is the study of the impacts of a single source, such as a new coal-fired power station, on a single critical atmospheric component, such as precipitation acidification. More complex assessments address the question of aggregate impacts across different kinds of sources. A contemporary example is the study of the net impact on Earth's thermal radiation budget caused by chemical perturbations due to fossil fuel combustion, biomass burning, land-use changes, and industrialization. An alternative approach, especially useful for the purposes of policy and management, is to assess the impacts of a single source on several critical environmental properties. The coal combustion study noted above would fall into this category if the impacts were assessed not only on acidification, but also on photochemical oxidant production, materials corrosion, visibility degradation, heavy metals emissions, etc. If desired, the columns could be summed in some way (such as assigning numbers to potential importance ratings) to give the net impact of the ensemble of sources on each critical property. Similarly, the rows could be summed in some way to give the net effect of each source on the ensemble of properties.

48 Chapter 4 Impact and Interpretation Analysis

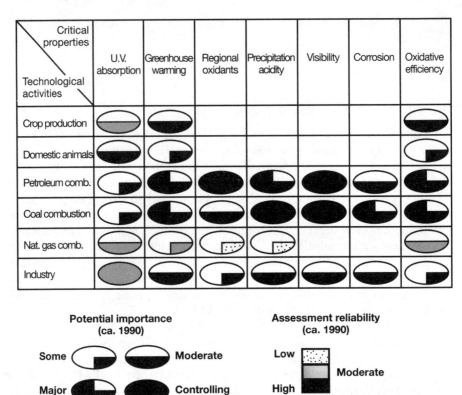

FIGURE 4.1 An initial ensemble assessment of impacts on the global atmosphere. Critical atmospheric properties are listed as the column headings of the matrix. The sources of disturbances to these properties are listed as row headings. Cell entries assess the relative impact of each source on each component and the relative scientific certainty of the assessment.

Figure 4.1 shows that the sources of most general concern for atmospheric impacts are fossil fuel combustion (especially coal and petroleum) and industrial processes (obviously the sum of many disparate activities). Emissions from agricultural operations, especially CH_4 from rice paddies and cattle and N_2O from fertilizer application, also have significant effects on climate and environment. When contemplating these source-impact assessments, it is useful to consider some of the differing attributes of the sources. Food production through the growing of crops and the raising of animals is potentially sustainable, but current operations require very large resource inputs. Sustainability would thus demand substantial and far-reaching changes in approach. The combustion of fossil fuels, necessarily preceded by extraction and purification, involves little transformation and is purely dissipative in nature. In contrast, industrial operations exist for the purpose of transforming raw materials, and are potentially nondissipative if sustainable energy supplies are assumed.

Among the most troublesome interactions between development and environment are those that involve cumulative impacts. In general, cumulative impacts become important when sources of perturbation to the environment are grouped sufficiently closely in space or time that they exceed the natural system's capability to remove or dissipate the resultant disturbance. For atmospheric perturbations, for example, the basic data required to structure such assessments are the characteristic time and space scales of the atmospheric constituents and development activities. For example, gases with very long lifetimes accumulate over decades to centuries around the world as a whole. Today's perturbations to those gases will still be affecting the atmosphere decades or centuries hence, and perturbations occurring anywhere in the world will affect the atmosphere everywhere in the world. Long-lived emittants tend to be radiatively active, thus giving the "greenhouse" syndrome its long-term, global-scale character. At the other extreme, heavy hydrocarbons and coarse particles, being short lived, drop out of the atmosphere in a matter of hours, normally traveling a few hundred kilometers or less from their sources. The atmospheric properties of visibility reduction and photochemical oxidant formation associated with these chemicals thus take on their acute, relatively local or regional character. Species with moderate atmospheric lifetimes include gases associated with the acidification of precipitation and fine particles associated with visibility decreases, each with characteristic scales of a few days and a couple of thousand kilometers.

This scale diversity is reflected in Figure 4.2, which reflects stressor interaction with water, soil, and other environmental properties. The figure contains several messages. One is that industrial activities play at least a small role in all the impacts, unlike other sources. For human organism damage, industry plays the major role. For impacts on species diversity, industrial activities are minor compared with crop production (and consequent forest clearing) as significant contributors. The same is true of soil productivity loss. In the case of groundwater quality, industrial activity and food production can each be important, with the dominant influence varying from location to location. The industrial system (defined in the very broad sense to include consumers and their activities) can be held predominantly responsible for the depletion of landfill capacity.

Industry is a major factor in the case of materials extraction, but not as large as the removal of fossil fuel resources for the production of energy. The use of energy thus bears two burdens: one as a producer of carbon dioxide, the principal anthropogenic greenhouse gas, the other as a major negative factor on local ecosystems through the impacts that accompany resource extraction.

4.3 INDUSTRIAL PRIORITIZATION: THE NETHERLANDS VNCI SYSTEM

Using the goal of the management of individual substances in industry as a focus, the Dutch government and the Association of the Dutch Chemical Industry (Vereniging van de Nederlandse Chemische Industrie, VNCI) have developed their own assessment methodology for studying the reductions of environmental impacts of a variety of chemical substances. The methodology proceeds in stages from generating options to prioritizing options to planning actions. The second stage, the prioritization of options, will be summarized here.

Prioritization begins with the development of an environmental profile for each option. This profile first identifies the "environmental themes" where implementing

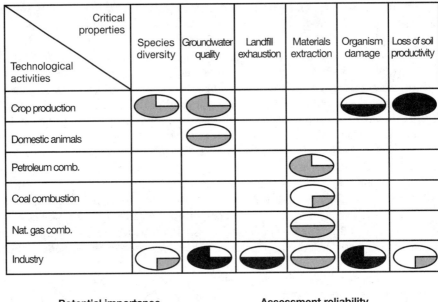

FIGURE 4.2 An initial ensemble assessment of impacts on global water and soil. The format is the same as that of Figure 4.1. While not prescriptive because of their qualitative nature, the displays nonetheless permit ready comparison of the multi-dimensional nature of the environmental interactions of different industrial options.

the option is likely to have an influence—global warming or disposal of waste, for example—and then assigns an index to the benefit or cost that will result. The next step is to construct an economic profile of each option. Elements typically captured at this stage include net changes in capital and operating costs, both positive and negative. Finally, the two profiles for each option are combined in a prioritization diagram, in which options can be compared either individually or in combination.

The inclusion of economic factors is an important and often overlooked aspect of LCA. In the case study in the previous chapter, it appeared that tallow required so much water that it should perhaps be rejected as a surfactant feedstock. Such a decision would overlook the fact, however, that beef cattle are primarily raised for meat, not for tallow. Utilizing the tallow is an efficient use of a low-value co-product that otherwise would require disposal. Hence, a broad economic perspective is often vital in producing a well-reasoned LCA result.

The Dutch procedure is best illustrated by example, as in a study aimed at reducing emissions of HCFC-22, a gas used in refrigeration and in the manufacture of fluoroplastics. Nine options for doing so were identified, and two were selected for detailed

Section 4.3   Industrial Priorization: The Netherlands VNCI System   51

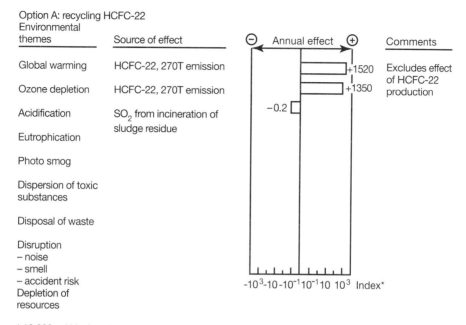

FIGURE 4.3   The environmental profile diagram for a proposed Netherlands program to recycle HCFC-22. The index is logarithmic and is based on 1% of total Netherlands HCFC-22 emissions being set to a value of 10,000. (Ministry of Housing, Physical Planning, and Environment, *Integrated Substance Chain Management*, Pub. VROM 91387/b/4-92, 's-Gravenhage, Netherlands, 1991.)

study: establishing a recycling option and improving equipment maintenance. The environmental profile for the first option is shown in Figure 4.3. The profile shows that beneficial impacts on Dutch contributions to global warming and ozone depletion would result. A slight negative impact is also forecast because of the related acidification that is expected from the incineration of the residue from the HCFC-22 recycling stream.

Figure 4.4 shows the economic profile for the HCFC-22 recycling option. Costs will be incurred in purchasing the equipment necessary for the recycling program and for the operating costs of performing the recycling. However, these costs are offset by reduced HCFC-22 purchases.

The environmental and economic profiles are combined on the prioritization diagram shown in Figure 4.5, and the ratings for the increased maintenance option are also given. The quantitative approach of the first two diagrams is retained in constructing this figure, but for convenience in assessment each axis is divided into low, medium, and high regions, each division indicating a factor of 10 difference from the adjacent one. The result of the study is that both options were judged to have a significant environmental yield in exchange for a slightly negative (Option B) or slightly positive (Option A) economic impact. The possible choice of implementing both options

Chapter 4 Impact and Interpretation Analysis

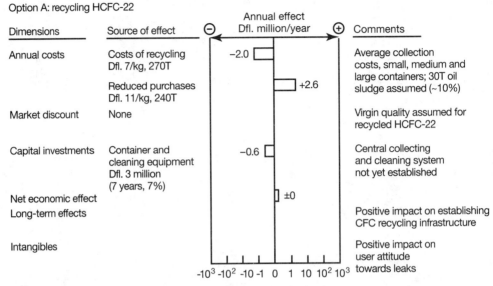

FIGURE 4.4 The economic profile diagram for a proposed Netherlands program to recycle HCFC-22. The index is monetary and logarithmic. (Ministry of Housing, Physical Planning, and Environment, *Integrated Substance Chain Management*, Pub. VROM 91387/b/4-92, 's-Gravenhage, Netherlands, 1991.)

results in a slightly larger environmental yield, still with only a slight negative economic impact.

The VNCI system is relatively new and not yet refined. Indeed, one can readily discern potential improvements. One characteristic that could be modified is that the system, as structured, rates all environmental themes equally (although the supporting documents hold out the potential for applying weighting factors). Another problematic characteristic is the logarithmic scale in the economic profile and prioritization diagram. Such an approach suggests that options differing significantly in cost might fall in the same "medium impact" region; it is doubtful that industrial manufacturing facility managers would regard an option costing three times that of another option as having essentially the same financial impact. Nonetheless, the VNCI system's inclusion of economic costs as an integral part of the analysis is a major favorable attribute. It is reasonable to anticipate that some version of this system could prove very useful for LCA prioritization.

4.4 INDUSTRIAL PRIORITIZATION: THE IVL/VOLVO EPS SYSTEM

An initial attempt to establish a more structural basis for assessing the environmental responsibility of an individual manufacturing process was that of Roger Sheldon of the

Section 4.4   Industrial Priorization: The IVL/VOLVO System   53

Prioritizing HCFC-22 options

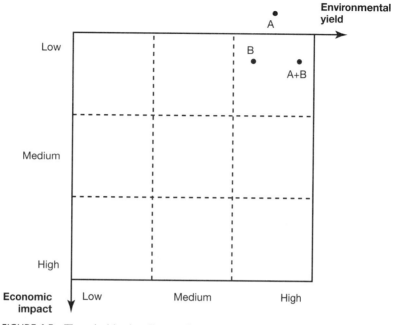

FIGURE 4.5 The prioritization diagram for a proposed Netherlands program to recycle HCFC-22. (Ministry of Housing, Physical Planning, and Environment, *Integrated Substance Chain Management*, Pub. VROM 91387/b/4-92, 's-Gravenhage, Netherlands, 1991.)

Delft (Netherlands) Institute of Technology, who proposed in the context of the synthesis of organic chemicals the *atom utilization concept* (AU), which is calculated by dividing the molecular weight of the desired product by that of the sum total of all products and residues produced. Enlarging on this concept, and realizing that the nature of the residue is important as well as its amount, he proposed the environmental quotient (EQ), given by

$$EQ = AU \times U \tag{4.1}$$

where   U is the "unfriendliness quotient," a measure of environmental hazard.

Sheldon offers no advice on how to assign unfriendliness quotients, but this difficult problem has not gone unaddressed. To begin to deal with something like it in a formal way, the Swedish Environmental Research Institute (IVL) and the Volvo Car Corporation have developed an analytic tool called the *Environment Priority Strategies for Product Design (EPS)* system. The goal of the EPS system is to enable product designers to select components and subassemblies that minimize environmental impact. Analytically, the EPS system is quite straightforward, though detailed. An environmental index is assigned to each type of material used in automobile manufacture. Different units of the index account for the environmental impact of this material during product manufacture, use, and disposal. The three life-cycle stage components are summed to obtain the overall index for a material, expressed in environmental load units (ELUs) per kilogram of material used (ELU/kg). The units may sometimes vary, however. For example, the index for a paint used on the car's exterior would be expressed in ELU/m^2.

When calculating the components of the environmental index, the following factors are included:

- *Value:* The willingness to pay to avoid the unit effect.
- *Distribution:* The size or composition of the affected environmental area.
- *Frequency or intensity:* Extent of the impact in the affected environmental area.
- *Durability:* Persistence of the impact.
- *Contribution:* Significance of impact from 1 kg of material in relation to total effect.

These factors should be calculated by a team of environmental scientists, ecologists, materials specialists, and economists to obtain environmental indices for every applicable raw material and energy source (together with their associated pollutant emissions). A selection of the results is given in Table 4.1, in which a few features are of particular interest. One is the very high values for platinum and rhenium in the raw materials listings, a result of the extremely low concentrations in Earth's crust of these two metals and the specific EPS method by which resources are valued. The use of CFC-11 is given a high environmental index as well because of its effects on stratospheric ozone and global warming. Finally, the assumption is made that the metals are emitted in a mobilizable form (i.e., they are in a chemical form that can interact with biological organisms). To the extent that an inert form is emitted, the environmental index may need to be revised.

Given an agreed upon set of environmental indices, those indices are multiplied by materials uses and process parameters (in the appropriate proportions) to obtain

TABLE 4.1 A Selection of Environmental Indices (units: ELU/kg)[*]

Raw Materials		Emissions–Air:	
Co	76	CO_2	0.09
Cr	8.8	CO	0.27
Fe	0.09	NO_X	0.22
Mn	0.97	N_2O	7.0
Mo	1500	SO_X	0.10
Ni	24.3	CFC-11	300
Pb	180	CH_4	1.0
Pt	350,000		
Rh	1,800,000	Emissions–Water:	
Sn	1200	Nitrogen	0.1
V	12	Phosphorus	0.3

[*]Steen, B., and S. Ryding, *The EPA Enviro-Accounting Method: An Application of Environmental Accounting Principles for Evaluation and Valuation of Environmental Impact in Product Design*, Swedish Environmental Research Institute (IVL), Stockholm, Sweden, 1992.

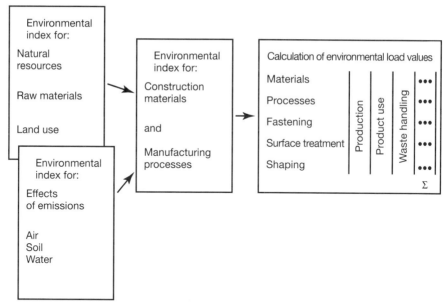

FIGURE 4.6 An overview of the EPS system showing how the calculation of summed environmental load values proceeds. (Reproduced with permission from B. Steen, and S. Ryding, *The EPA Enviro-Accounting Method: An Application of Environmental Accounting Principles for Evaluation and Valuation of Environmental Impact in Product Design.* Copyright 1992 by Swedish Environmental Research Institute [IVL].)

environmental load units for processes and finished products. The entire procedure is schematized in Figure 4.6. Notice that the process includes the effects of materials extraction, emissions, and manufacturing, and follows the product in its various aspects through the entire life cycle.

As an example of the use of the EPS system, consider the problem of choosing the more environmentally responsible material to use in fabricating the front end of an automobile. As shown in Figure 4.7, two options are available: galvanized steel and polymer composite (glass-mat thermoplastic, or GMT). The front ends are assumed to be of comparable durability, although differing durabilities could potentially be incorporated into EPS.

Based on the amount of each material required, the environmental indices are used to calculate environmental load values at each stage of the product life cycle. Table 4.2 illustrates the total life cycle ELUs for the two front ends. All environmental impacts, from the energy required to produce a material to the energy recovered from incineration or reuse at the end of product life, are incorporated into the ELU calculation. To put the table in the LCA perspective, the kg columns are the results of LCA stage one, the ELU/kg columns are the environmental indices, and the ELU columns are the results of LCA stage two, the latter being given by

$$\text{ELU} = \sum_i \sum_s (\text{ELU/kg})_i \, M_{i,s} \qquad (4.2)$$

Which front end is more environmentally sound?

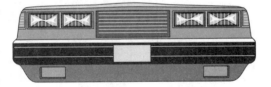

GMT – composite
Material consumption: 4.0 kg
(0.3 kg scrap)
Component weight: 3.7 kg

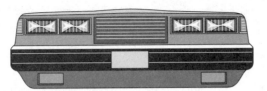

Galvanized steel
Material consumption: 9.0 kg
(3.0 kg scrap)
Component weight: 6.0 kg
Painted area: 0.6 m²

FIGURE 4.7 Design options for automotive front-end pieces. (Courtesy of I. Horkeby, Volvo Car Corporation.)

where   i indicates the type of material, s the life stage, and M is the mass of material i at life stage s.

There are several features of interest in the results. One is that the steel unit has a larger materials impact during manufacturing, but is so conveniently reusable that its end of life ELU is lower than that of the composite. However, the steel front end is much heavier than the composite unit, and that factor results in much higher environmental loads during product use. The overall result is one that was not intuitively obvious: The polymer composite front end is the better choice in terms of environmental impacts during manufacture, the steel unit the better choice in terms of recyclability, and the polymer composite unit the better overall choice because of lower impacts during product use. Attempting to make the decision on the basis of an analysis of only part of the product life cycle would result in an incompletely guided and potentially incorrect decision.

The EPS system is currently being refined and implemented by several organizations and researchers. Corporations in Sweden and elsewhere have expressed great interest in developing EPS systems that are specific to their products and manufacturing procedures. The EPS's greatest strength is its flexibility; raw materials, processes, and energy uses can be added easily. If a manufacturing process becomes more efficient, all products that utilize the process will automatically possess upgraded ELUs. Perhaps its greatest weakness is its need to quantify uncertain data and compare unlike risks, in the process making assumptions that gloss over serious value and equity issues.

TABLE 4.2 Calculation of Environmental Load Values for Automobile Front Ends[*]

Materials & Processes	Production			Product Use[‡]			Waste						Total ELU
							Incineration			Reuse			
	ELU/kg	kg	ELU	ELU/kg	kg	ELU	ELU/kg	kg	ELU	ELU/kg	kg	ELU	
GMT-composite													
Production:													
GMT material	0.58	4.0	2.32										2.32
Reused production scrap	−0.58	0.3	−0.17										−0.17
Compression molding	0.03	4.0	0.12										0.12
Product Use:													
Petrol				0.82	29.6	24.27							24.27
Recycling:													
GMT material							−0.21	3.7	−0.78				−0.78
Total sum			2.27			24.27			−0.78				25.76
Galvanized steel													
Production:													
Steel material	0.98	9.0	8.82										8.82
Steel stamping	0.06	9.0	0.54										0.54
Reused production scrap	−0.92	3.0	−2.76										−2.76
Spot welding (spots)	0.004	48	0.19										0.19
Painting (m²)	0.01	0.6	0.02										0.02
Product Use:													
Petrol				0.82	48.0	39.36							39.36
Recycling:													
Steel material										−0.92	6.0	−5.52	−5.52
Total sum			6.81			39.36						−5.52	40.65

[*]S. Ryding, B. Steen, A. Wenblad, and R. Karlson, The EPS System—A Life Cycle Concept for Cleaner Technology and Product Development Strategies, and Design for the Environment, paper presented at EPA Workshop on Idneitfying a Framework for Human Health and Environmental Risk Ranking, Washington, DC, June 30–July 1, 1993.
[‡]The ELU/kg figure is based on one year of use. For the automobile, an eight-year life is assumed, hence the second product use entry is eight times the actual weight.

Chapter 4   Impact and Interpretation Analysis

4.5 INDUSTRIAL PRIORITIZATION: THE NETHERLANDS NSAEL METHOD

The sequence of steps involved in LCA impact analysis is addressed more explicitly than in the EPS system by the *NSAEL method*, a product of environmental scientists in the Netherlands. To demonstrate the procedure, consider an assessment that has been made of the environmental impacts of window frames designed and manufactured by four different corporations. A selection of the emissions to air and water chargeable to the window frames (i.e., the results of the inventory analysis) is given in Table 4.3.

It clearly makes little sense to merely add the columns in Table 4.3, because the impacts of the emittants are quite different and because some emittants have impacts on more than one environmental concern. Accordingly, the first step in LCA impact analysis, classification, must be performed. For example, NO_x and butanone both influence human toxicity, CFCs and HCFC-22 both influence ozone depletion, and so on. In the characterization step, the emissions pertaining to a common impact are placed on the same quantitative scale. The impact score for environmental concern is then given by

$$I_{C,e} = \sum_x C_{x,e} \cdot M_x \qquad (4.3)$$

$C_{x,e}$ is the characterization factor for emittant x, and M_x is the mass of emittant x per window frame. In the case of ozone depletion, for example, computer models have calculated the depletion to be expected per molecule of impacting substance emitted. The results are related to the impact of a molecule of CFC-11, whose ozone depletion potential (ODP) is assigned as 1. For HCFC-22, that turns out to be 0.05. Assuming the emittants in Table 4.3 constitute a complete list (although actually many other emittants were catalogued), the characterized ozone depletion impact for window frame D is given by

TABLE 4.3   Emissions Chargeable to the Production of Window Frames*

Emitted Species	Window Frame			
	A	B	C	D
Air emissions (kg/frame)				
CO_2	220	1400	190	360
CFC-11	<0.0001	<0.0001	0.0004	0.002
HCFC-22	0	0	0	0.1
NO_x	0.01	0.005	0	0.006
Water emissions (kg/frame)				
Cl	0.3	0.2	3.2	0.3
NH_3	0.07	0.06	0.05	0.6
P	0	0	0	0.2
Aromatic HCs	0.004	0.002	0	<0.001

*Abstracted from a list of more than ninety emittants in J.G.M. Kortman, E.W. Lindeijer, H. Sas, and M. Sprengers, *Towards a Single Indicator for Emissions—An Exercise in Aggregating Environmental Effects*, Interfaculty Dept. of Environmental Science, Univ. of Amsterdam, The Netherlands, March, 1994.

Section 4.5   Industrial Priorization: The Netherlands NSAEL System   59

$$I_{C\,(\text{ozone depletion})} = \sum_{x} \text{ODP}_x \cdot M_x = 0.002 \cdot 1 + 0.1 \cdot 0.05 = 0.007 \qquad (4.4)$$

In principle, characterized impacts can be derived for other environmental concerns such as human organism risk and acidification, although this has often proven difficult in practice. The window frame study considered five concerns (termed "environmental aspects"), computed their characterized impacts, and assigned the value for the window frame with the highest score for each concern to 100. The results are shown in Figure 4.8, where, at this point in the impact analysis, all concerns are treated equally.

It is important to realize that because the characterization factors for the concerns are derived quite differently, adding the characterized ratings for the five concerns is inappropriate. Rather, one is limited to noting that, for example, window frame A receives the best (i.e., lowest) score on three of the five concerns and ties with window frame B for the best score on a fourth.

The next LCA step is localization, in which the characterized impacts are referenced to the region in which the impacts occur. In the first phase of localization, the characterized impacts are related to the magnitudes of the total characterized impacts for the region (the Netherlands as a whole), i.e.,

$$I_{N_1(e)} = \frac{I_{C,e}}{I_{C,e}\,(\text{region})} \qquad (4.5)$$

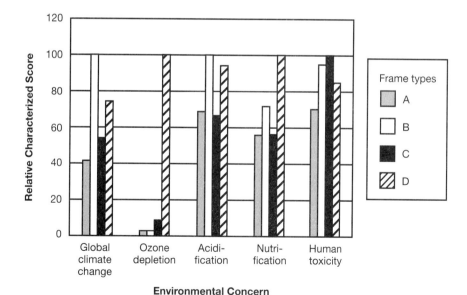

FIGURE 4.8   Characterized impacts of the manufacture of four different window frames on five environmental concerns. (Adapted from J.G.M. Kortman, E.W. Lindeijer, H. Sas, and M. Sprengers, *Towards a Single Indicator for Emissions—An Exercise in Aggregating Environmental Effects*, Interfaculty Dept. of Environmental Science, Univ. of Amsterdam, The Netherlands, March, 1994.)

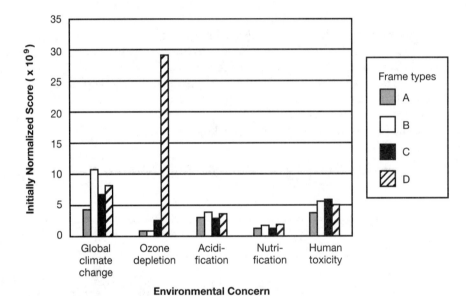

TABLE 4.4 Normalization Parameters for The Netherlands in the Window Frame LCA

Environmental concern	$I_{C,e}$ (region)	$N_2(e)$
Global climate change	2.5×10^{11}	1.6
Ozone depletion	2.3×10^5	4.0
Nutrification	1.1×10^9	2.5
Acidification	5.9×10^8	4.9
Human toxicity	9.4×10^8	0.6

FIGURE 4.9 Initially normalized impacts of the manufacture of four different window frames on five environmental concerns. (Adapted from J.G.M. Kortman, E.W. Lindeijer, H. Sas, and M. Sprengers, *Towards a Single Indicator for Emissions—An Exercise in Aggregating Environmental Effects*, Interfaculty Dept. of Environmental Science, Univ. of Amsterdam, The Netherlands, March, 1994.)

The values in the denominator are shown in Table 4.4, and the results of the calculation are plotted in Figure 4.9. The $I_{N_1(e)}$ values provide somewhat more perspective than the $I_{C,e}$ values because the former show the contribution to each environmental concern of manufacturing a window frame relative to the contributions from all other activities within the region.

The second phase of localization involves assessing the assimilative capacity of the region for the environmental concern in question, a phase that obviously requires the availability of environmental data at a spatial scale consistent with that of the LCA. The approach taken in the NSAEL method is to determine for each concern the No Significant Adverse Effect Level, that is, the level of impact at which structural changes to the ecosystem in question do not occur. For example, in the case of acidification, the chosen target effect is the onset of systematic damage to forests, for which the NSAEL level has been determined to occur at an input flux of acid equivalents to the forest floor of about 1400 $[H^+]$ ha^{-1} yr^{-1}, where $[H^+]$ indicates the potential formation of one mole of hydrogen ions. These "acidification equivalents" can be related to emissions of SO_2, NO_x, and NH_3, which, for the Netherlands, amount to 3.4×10^{10} $[H^+]$ yr^{-1}, or 8200 $[H^+]$ ha^{-1} yr^{-1}. The assimilative factor in normalization is then given by

$$N_2(e) = \frac{E_A - E_{NSAEL}}{E_{NSAEL}} \tag{4.6}$$

with the provision that $N_2(e) = 0$ for $E_A \geq E_{NSAEL}$, and, for the acidification example above,

$$N_2(\text{acidification}) = \frac{8200 - 1400}{1400} = 4.9 \tag{4.7}$$

The normalized impact is then given by

$$I_{N(e)} = I_{N_1(e)} \cdot N_2(e) \tag{4.8}$$

The initially normalized results from Figure 4.9 can thus be multiplied by the N_2 (acidification) values (see Table 4.4) to give impacts weighted by the assimilative capacity of the ecosystem into which the emissions occur; these are the fully normalized impacts.

If full normalization is the final step performed in the impact assessment, as is often the case, the total environmental impact of the product is computed by summing over all environmental concerns.

$$I_N = \sum_e I_{N(e)} \tag{4.9}$$

The final, if seldom attempted, step in LCA impact analysis is valuation, in which the importance of different impacts on the individual environmental concerns are weighted according to some societal consensus, i.e., by the use of weighting factors.

$$I_{V_e} = I_{N(e)} \cdot V_e \tag{4.10}$$

The overall valuated impact rating for a product is then given by

$$I_V = \sum_e I_{V(e)} \tag{4.11}$$

In the case of the window frames, valuation was not attempted and Eq. (4.9) was used to produce the results shown in Figure 4.10. On the basis of the environmental concerns included in the study, window frame D clearly has a higher total impact than do the other three, whose total impacts are quite similar.

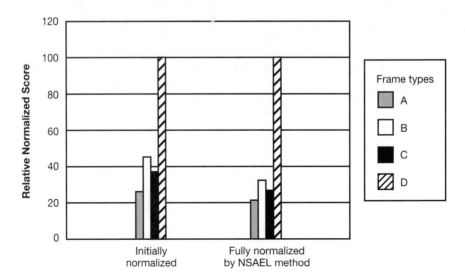

FIGURE 4.10 Fully normalized impacts of the manufacture of four different window frames on five environmental concerns. (Adapted from J.G.M. Kortman, E.W. Lindeijer, H. Sas, and M. Sprengers, *Towards a Single Indicator for Emissions - An Exercise in Aggregating Environmental Effects*, Interfaculty Dept. of Environmental Science, Univ. of Amsterdam, The Netherlands, March, 1994.)

4.6 INTERPRETATION ANALYSIS

4.6.1 Prioritization Tables

After the impact analysis is completed, to the greater or lesser degree possible, one proceeds to the final LCA component: *interpretation analysis*, in which the results of the inventory and impact analyses are used to derive perspectives on the system, or, better yet, recommendations for actions to minimize the detrimental environmental attributes of the system under study. The environmental responsibility of an assessed product can usually be substantially improved by adopting most or all of the recommendations made in the assessment report. Implementation of all the recommendations made in response to an LCA/SLCA cannot, however, be accomplished simultaneously. Prioritization is thus useful, and in order to prioritize the recommendations one should consider more than just DFE-related characteristics. Among the straightforward and efficient ways to establish prioritization is to rank each recommendation on a "+/−"scale ("++" being the most desirable score and "−−" being the least desirable score) according to the following considerations:

- *Technical feasibility:* Estimates the technical difficulty of implementing a particular recommendation; "++" means the recommendation presents no technical challenges and is therefore very easy to implement.
- *Environmental improvement:* Judges to what extent implementation of a recommendation will respond to an important environmental impact, the situation

being evaluated on both a scientific and social basis; "++" means inclusion of a particular DFE attribute will strongly support desirable environmental initiatives.
- *Economic benefit:* Evaluates the net cost of implementing a particular recommendation; "++" means the product would cost less if recommendation is incorporated. Here the total life-cycle cost is considered. For example, some piece parts may cost more due to DFE constraints but will also yield a higher residual value when the equipment is retired.
- *CVA impact:* Accounts for the customer-perceived value added (CVA) resulting from implementing a particular recommendation; "++" means the DFE attribute has a very high perceived value. This parameter is completed from the perspective of the most environmentally conscious customer.
- *Product management:* Estimates the production schedule impact or other product management result of implementing a particular recommendation; "++" means adoption of the recommendation would reduce the amount of time required to develop/implement the product; +/− means it would have no significant effect.

An example of the prioritization of recommendations generated in connection with a DFE assessment for an electronic product is given in Table 4.5. (Details of the reasons for the recommendations themselves are given in the Industrial Ecology textbook cited in the Preface.) The individual scores were assigned by the DFE assessor

TABLE 4.5 A Prioritization Table for DFE Recommendations

Recommendations	Life stage	Technical feasibility	Environmental sensitivity	Economic impact	CVA impact	Production management	Total
Manufacturing:							
Recycled metal specs.	L1.1	++	++	+/−	+	+/−	15
Packaging diversity-inflow	L2.1	++	+	+/−	+/−	+/−	13
Packaging diversity-outflow	L3.1	++	+	+/−	+	+/−	14
Reusable ship. containers	L3.2	++	+	+/−	+	+/−	15
Solder bath N$_2$ inerting	L2.2	++	++	−	+/−	−	12
Design:							
Avoid chromate	L1.2(5)	+	+	+/−	+/−	+/−	12
Less plastic diversity	L5.1	+/−	+	+/−	+	−	11
Mark plastic parts	L5.2	++	++	+/−	+	+/−	15
Management:							
On-line information	L4.1	++	+	−	+	−	12
Battery recovery	L4.2	++	++	−	++	+/−	15

Symbol	Value	Points
++	Very good/high	4
+	Good/high	3
+/−	Moderate, average	2
−	Little/bad	1
− −	Very little/bad	0

and the recommendations were then sorted, in order of decreasing overall value to the manufacturing organization, in each of the three categories: Manufacturing, Design, and Management.

4.6.2 Prioritization Diagrams

4.6.2.1 The Action-Agent Prioritization Diagram

Although the prioritization table is helpful in developing additional supporting information relative to LCA recommendations, its extensiveness may make the most significant information difficult to extract readily, particularly if the number of recommendations is larger than shown here. An alternate display of the information is with a *prioritization diagram*, as shown in Figure 4.11, which is a variation of the widely used Pareto plot. The first step in constructing the diagram is to normalize the assessment sum of Table 4.5 by reducing each sum by 10; the philosophy is that the maximum score is 20 and a score at or below 10 may be regarded as pertaining to a recommendation that would produce little net benefit. The practical benefit of the adjustment is to make it easier to distinguish between and choose among the more highly rated recommendations. The adjusted prioritization sums are plotted in three groups, each group representing recommendations that would need to be carried out by specific "action agents": manufacturing engineers, design engineers, or management personnel.

The highest priority recommendations are quickly distinguished from those of lower priority in Figure 4.11. In the manufacturing area, two actions have the highest priority rating: (1) Specify that major metal parts contain recycled content, and (2) use reusable shipping containers for modules and components. Several other actions listed

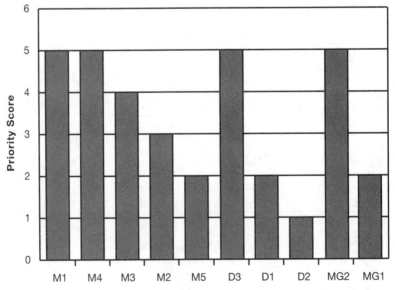

FIGURE 4.11 An action-agent prioritization diagram of the recommendations from the streamlined life-cycle assessment of a telecommunications product.

in the table are rated high (though not highest) in priority; accomplishing these would also be well justified. The economic impact for all these actions is small to negligible.

In the design area, the recommendation that stands out is to mark the major plastic parts with ISO symbols. The other recommendation is also of merit.

For management, one priority action is identified: The development of a program to efficiently take back discharged batteries from the field. Less important is to provide online maintenance and upgrading information to customers.

4.6.2.2 The Life-Stage Prioritization Diagram

An alternative display of the prioritized recommendations is provided by a *life-stage diagram*. As with the action-agent diagram, the basic information is taken from Table 4.5 and normalized. The recommendations are then divided into five groups, one for each life stage: premanufacture, manufacturer, product delivery, product use, and end of life. If a recommendation pertains to more than one life stage, it is included in each life stage group to which it pertains. The result for the telecommunications product example is shown in Figure 4.12.

The life-stage diagram provides a different perspective on the recommendations, one that varies in time and space rather than in the action agents. The environmental aspects of the manufacturing stage, for example, are seen to rate rather highly, as the priority scores of the applicable recommendations are low. In contrast, the end-of-life stage has recommendations with higher priority scores. Attention is also indicated for the product use stage. The latter two stages are under the direct control of product designers. The premanufacture stage merits activity that requires the participation of the procurement organization in working with suppliers.

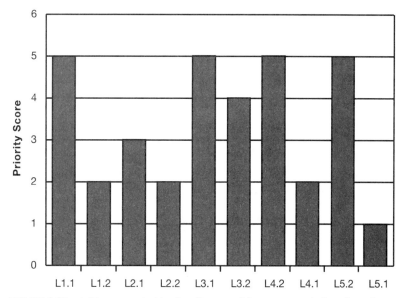

FIGURE 4.12 A life-stage prioritization diagram of the recommendations from the streamlined life-cycle assessment of a telecommunications product.

4.6.3 Aspects of Interpretation

The approach to improvement analysis outlined in this chapter basically consists of two parts: deriving a set of recommendations from earlier LCA stages and then prioritizing those recommendations from the standpoint of both environmental and non-environmental factors. Comprehensive LCAs, streamlined LCAs, or even scoping studies may be used as the information source from which the recommendations are derived. Tables and graphical displays are used as tools for communicating the results.

Prioritization of recommendations is an essential step in improvement analysis. The method used here to prioritize the LCA recommendations is not necessarily the only one that is feasible, and it may seem inappropriate to some to give the same weight to economic impacts and project management implications as is given to environmental consequences. Such an approach is probably a realistic one, however, and is presented as a utilitarian and widely understood way in which to accomplish a prioritization that will receive general concurrence.

It is interesting to examine how this improvement assessment process differs as a function of whether those using the tools are industrial firms, policy makers, or interested third parties. The first step, deriving recommendations, could, in principle, provide identical results no matter which type of agency is performing the assessment. (In practice, the most useful recommendations will probably come from the firms closest to the product or process under study.) The second step, prioritization, seems likely to be much less uniform, as it involves information on customer preferences and internal and external cost; these items are unlikely to be available to non-industrial parties. From a pragmatic standpoint, therefore, LCA improvement tools are probably useful primarily within the organization whose operations are under assessment.

Although the examples in this chapter have treated the improvement analysis of products, a similar approach would be effective for processes, facilities, service industries, and infrastructure. Life-stage diagrams and SCLA approaches to these other applications have been derived and, in some cases, implemented.

Improvement analysis is clearly in its formative stages, and the tools presented here will doubtless evolve over time. A formal improvement analysis is crucial to an LCA if benefits are to be obtained, however, and this chapter begins a dialog on the development of techniques needed to make that implementation occur.

FURTHER READING

Boguski, T.K., R. G. Hunt, J.M. Cholakis, and W. E. Franklin, LCA methodology, in *Environmental Life-Cycle Assessment*, M.A. Curran, Ed. (New York: McGraw-Hill, 1996), pp. 2.1–2.37.

Edmunds, R.A., *The Prentice Hall Guide to Expert Systems* (Englewood Cliffs, NJ: Prentice-Hall, 1988).

Fava, J., et al., *A Conceptual Framework for Life-Cycle Impact Assessment* (Pensacola, FL: Society for Environmental Toxicology and Chemistry, 1993).

Graedel, T.E., Industrial ecology: Definition and implementation, in *Industrial Ecology and Global Change*, R. Socolow, C. Andrews, F. Berkhout, and V. Thomas, Eds. (Cambridge, U.K.: Cambridge Univ. Press, 1994) pp. 23–41.

Guinee, J.R., R. Heijungs, H.A. Udo de Haes, and G. Huppes, Quantitative life cycle assessment of products: 2. Classification, valuation, and improvement analysis, *Journal of Cleaner Production, 1* (1993) 81–91.

Kortman, J.G.M., E.W. Lindeijer, H. Sas, and M. Sprengers, *Towards a Single Indicator for Emissions—An Exercise in Aggregating Environmental Effects* (Interfaculty Department of Environmental Science, Univ. of Amsterdam, March, 1994).

Lindfors, L. G., et al., *Impact Assessment*, Technical Report No. 10, Tema Nord 1995:503 (Copenhagen: Nordic Council of Ministers, 1995)

Ministry of Housing, Physical Planning, and Environment, *Integrated Substance Chain Management*, Pub. VROM 91387/b/4-92, ('s-Gravenhage, Netherlands, 1991).

Steen, B., and S. Ryding, The EPA Enviro-Accounting Method: An Application of Environmental Accounting Principles for Evaluation and Valuation of Environmental Impact in Product Design (Stockholm: Swedish Environmental Research Institute (IVL), 1992).

Udo de Haes, H.A., Ed., *Towards a Methodology for Life Cycle Impact Assessment*, ISBN 90-5607-005-3 (Brussels: Society of Environmental Toxicology and Chemistry-Europe, 1996).

EXERCISES

4.1. Using the U.S. EPA Science Advisory Committee lists of risks from Chapter 1, create your own risk prioritization list. Explain and defend your choices, in each case differentiating between scientific and technical assessments on the one hand and values and ethical judgments on the other.

4.2. Acid rain is generated by the following process (among others): coal containing sulfur is burned in a fossil fuel power plant and some of the sulfur, now oxidized to sulfur dioxide, escapes the scrubber and enters the atmosphere. There it is transformed to sulfuric acid, which is adsorbed into raindrops, making the rain acidic. When the acidified rain falls, it depletes or destroys fish and other freshwater biota and may cause lower yields from agricultural crops. Identify the stressor or stressors mentioned and discuss the potential and actual environmental impacts.

4.3. It has been suggested (K. Saur, Life-cycle interpretation—A brand new perspective?, *International Journal of Life-Cycle Assessment, 2* (1997) 8–10) that several activities should precede the prioritization step of Table 4.5: Identification of major burdens and impacts, an error assessment, and a sensitivity analysis. For the EPS study of Table 4.2, attempt to identify those burdens and impacts qualitatively, suggest which of the tabular entries might be most subject to error and why, and to which errors the analysis would be most sensitive. Discuss your reasoning.

4.4. (a) Suppose that global warming is thought less likely to occur than had previously been assumed, and that, as a result, the ELU/kg for product use is lowered to 0.6. What effect does this have on the comparative ratings of the two front ends in Figure 4.7? (b) A new high-strength honeycomb steel has been developed and is being considered for use in automobile front ends. Rather than the steel front end weighing 6.0 kg, a satisfactory front end weighing only 4.0 kg can be formed from 6.0 kg of the new steel. Because the new steel's improved properties, which are due to added trace alloying elements, have negligible effects on processing or recycling, the same ELU/kg assessments apply to those stages (Table 4.2). Compute the ELU values for the new front end and compare them with the two options in the table. (c) Assume that the global warming revision of part (a) occurs as well as the availability of the new honeycomb steel front end. What effect do these two changes taken together have on the relative impact results? (d) Suppose that the shortage of petroleum (used as a feedstock for the manufacture of plastic composites) became so great that the ELU/kg of the composite materials is set to 1.90 and that the honeycomb steel front end is available. What effect do these two changes taken together have on the relative impact results? (e) What are the messages to designers implied in the analyses in the earlier parts of this exercise?

CHAPTER 5 | Evaluating the LCA Process

The LCA process described in the previous chapters appears relatively rigorous. The implication is that the results of the process lead ineluctably to environmentally preferable decisions. In practice, however, the total number of comprehensive LCAs has not been large, and many of their results have been regarded with skepticism. In this context, it is useful to evaluate some of the features of LCAs and to summarize some of the results that have been obtained. This information then permits an assessment to be made of the overall utility of the LCA process.

5.1 SYSTEM BOUNDARY CONSIDERATIONS

The scoping process in LCA, that is, defining what products or product families are to be examined and by what means the associated systems are to be defined and described, in turn defines how comprehensive the LCA will be. It is straightforward to include in the assessment the environmental consequences of emissions from a product manufacturing process, and only slightly less straightforward to treat the emissions attributable to energy consumption by the product during its useful lifetime. If the raw materials are purchased from an extractor or processor of ore, however, should the impacts of the extraction process be considered by the purchaser? If so, what about the materials and energy used in the extraction equipment? What about the process of transporting the ore from mine to smelter? How about smelter impacts? And, in the broadest view, if impacts attributable to extraction machines and processing machinery are taken into account, should one also consider the resources needed to sustain the workers that operate the machinery needed for extraction, transportation, and smelting?

In addition to questions about how far back and how far forward one should look, the analyst must also decide whether to include or exclude alternative approaches. An assessment of dry-cleaning of clothing, for example, might include the option to consider different cleaning solvents. One could also consider alternative ways of providing the service, such as making clothing with fibers that do not require dry cleaning or that need cleaning less frequently.

How much attention should be given to materials that are minor constituents? Sometimes components or materials whose fraction of total mass is below some arbitrary cutoff, say 5 or 10 percent, may be excluded in the interest of analytical tractability. This simplification is often defensible, but the unwary can be trapped by it if, for example, a minor but highly toxic material is contained in the product or emitted during the production of the product.

A final difficulty in defining the boundaries for inventory and impact analysis, and one that often encourages dispute, is that analyses of the same product but with different boundary choices often arrive at different results. A well-known example is a study of refrigeration in which energy use was treated as a penalty but the resultant useful heat was ignored as a benefit; this situation arose because the refrigeration equipment itself was within the boundaries of the assessment but the use of energy for the building in which the equipment was contained was outside the boundaries. In other examples, the inclusion or exclusion of the environmental impacts of transporting products or of generating needed energy determined whether impacts of manufacturing were important or not.

5.2 FUNCTIONAL UNIT CONSIDERATIONS

As has been seen, in order to compare the environmental impacts of one type of product with another, it is necessary to define the *functional unit*, that is, the amount of a product, material, or service that is to be assessed. The basic idea is that equivalent functional capabilities should be compared rather than arbitrary capabilities selected merely because of analytical convenience. The concept is straightforward when comparing a glass bottle to an aluminum can: It is a product holding a fixed amount (say 0.3 liters) of liquid. It is less obvious when comparing more complex products. For example, the analyst should not compare one washing machine directly with another, but rather should operate on the basis of the volume of clothing that each can wash. And, because a machine with a smaller internal washing volume may be designed with heavier materials in order to last longer, the proper functional unit might not be cubic meters washed, but perhaps cubic meters of capacity times years of expected operation. For automobiles, one might compare on the basis of impacts per kilometer of travel over anticipated vehicle lifetime rather than impacts per vehicle over a fixed short time period.

Hence, the functional unit must be chosen based on the job the product is designed to perform and how that job is carried out. The functional unit choice will inevitably involve the capabilities of the product: job capacity, efficiency, and expected length of service, and thus will incorporate assumptions that cannot necessarily be defended. These uncertainties provide tenuous underpinnings for comprehensive LCAs.

5.3 DATA LIMITATION CONSIDERATIONS

Those who have performed one or more LCAs are in agreement that the availability and accuracy of the data are always at least partly unsatisfactory. The degree of difficulty is strongly related to the boundary choices made for the assessment. If the boundaries are tightly constrained, the data may be readily available or at least easy to acquire. Very broad boundaries, in contrast, create substantial data requirements. If a product is being assessed, for example, and its raw material extraction impacts are included in the assessment boundaries, the assessor will need to know the actual or average levels of emissions, energy use, land conservation, and other activities of the raw material suppliers. Those data, which are never easy to acquire, become very difficult indeed in a world where goods trade rapidly across national and continental boundaries. Should one average emissions from mines in China, Chile, and Australia? Should a preference be expressed for raw materials sources on the basis of source environmental attributes? Should one try to anticipate technological improvements over time in the environmentally related activities of suppliers?

In acquiring relevant data, it is inevitable that data will be drawn from information compiled at different times or relating to materials produced over broadly different time periods. For short-lived products of modern technology, the data are likely not to have significant age distribution problems. For a building or an industrial process facility, however, one may be dealing with the technology, processes, and waste disposal practices of several different decades. The awkwardness can be especially difficult when comparing two or more products or services. An example used in Scandinavia is that of assessing electric power that is generated either by hydroelectric plants or by burning natural gas. In the former case, the concrete and steel of the plant and the aluminum and copper of the distribution system are decades old, while the piping of the natural gas supply system is relatively new. Does one use 1960s inventory data for one and 1980s inventory data for the other?

Chemicals and chemical processes present other problems. One problem is that there is no universal database containing toxicities and other environmental parameters for the many thousands of common industrial chemicals. Each assessor is thus forced into an independent fact-finding mission within the scientific literature. Another difficulty is that new chemicals constantly appear on the scene, with perhaps little or no environmentally related data accompanying them. An assessment that treats individual chemicals as well as more complex mixed products must thus make assumptions about toxicity, smog-forming potential, and the like. Finally, unless one knows the details of the specific operations employed by one's suppliers (knowledge unlikely in a world market with numerous potential suppliers), one must assume a certain level of technology and thus of associated environmental impacts for material suppliers without knowing who those suppliers may be. Because facilities using higher levels of technology tend to have substantially lower impacts than do those with low technological sophistication, this choice can significantly affect the outcome of an LCA, particularly as it relates to liquid residues, process exhaust gases, and combustion products.

Ultimately, those performing comprehensive LCAs need databases of the relevant compositions and environmental impacts of all materials in common industrial use. Similar databases would provide information on common industrial processes. At present, such databases are extremely limited. As a consequence, conscientious LCA assessors evaluating the same product can easily produce results that differ greatly as a consequence of the way in which information was generated to supplement readily available data.

5.4 IMPACT ASSESSMENT CONSIDERATIONS

5.4.1 Thresholds and Nonlinearities

A goal of LCA impact analysis, at least in principle, is to relate an environmental forcing or loading, that is, a flow of energy or materials chargeable to a product being

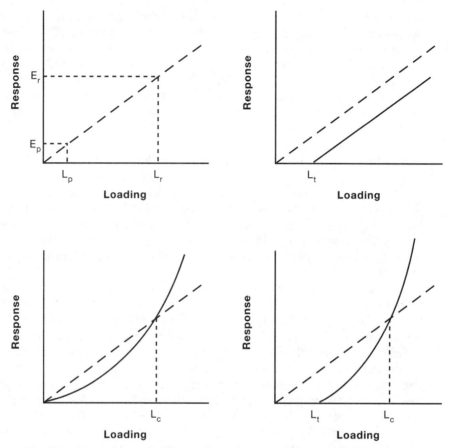

FIGURE 5.1 (a) A linear environmental loading and response relationship. (b) Environmental loading and response relationship for an environmental impact having a threshold at L_t. (c) Environmental loading and response relationship for an environmental impact having a nonlinear response that crosses the presumed linear relationship at L_c. (d) Environmental loading and response relationship for an environmental impact having both a threshold and a nonlinear response.

assessed, to the overall forcing or loading produced by all activities of society. Thus, the emissions of SO_2 chargeable to a product may be related to the sum of all emissions of SO_2 within a region. The product might then be charged with a proportionate share of the detrimental environmental impact due to sulfur-related acid rain. In Figure 5.1a, for example, if a regional loading L_r is determined to cause a regional effect E_r, a loading L_p charged to the product is considered responsible for effect $E_p = E_r (L_p/L_r)$.

Suppose, however, that the true relationship has a loading threshold below that at which no effect occurs, as shown in Figure 5.1b. Many processes resulting in damage to environmental systems or human organisms do indeed exhibit such a threshold; that is, a low level of exposure does not produce an effect but a higher level does. Thresholds occur because the receiving system is somehow buffered against the impact. For example, acidic precipitation falling on soils with high organic content has no effect as long as the acidic input does not exceed the buffering capacity. If the LCA impact assessment does not take the threshold into account, a loading less than L_t will produce a "false positive," a determination of a negative environmental impact where none exists.

Another difficulty occurs if the relationship between loading and effect is nonlinear, also a frequent occurrence. As shown in Figure 5.1c, loadings less than the "crossover loading" L_c will have their effects overestimated, while loadings more than L_c will have their effects underestimated.

Finally, if the loading-effect relationship is both nonlinear and has a threshold (Fig. 5.1d), opportunity exists for a false positive, an underestimation, or an overestimation, depending on the chargeable loading and the values of L_t and L_c.

5.4.2 Temporal Scales

Difficulties in LCA impact assessment are also known to arise from lack of consideration of temporal scales. For example, consider the patterns of emissions of volatile organic carbon compounds (VOCs) illustrated in Figure 5.2. VOCs are a main constituent of photochemical smog, and their emissions are highly regulated. Suppose that the emission rate is constant over time, as in the dash-dot pattern. In that case, the emissions will have a strong impact on local ozone production because the early morning VOCs will interact with automotive oxides of nitrogen NO_x emissions to generate ozone, the principal smog constituent. Alternatively, suppose an equivalent amount of VOC is emitted, but with a pattern like the dashed line pattern of Figure 5.2. In that case, because all emissions occur in the absence of sunlight, they have little or no effect on local ground-level ozone concentrations, and because of dispersion, little effect on distant ozone concentrations.

The difficulty is even more pronounced if there is a long time interval between action and consequences, as appears to be the situation with global climate change. The

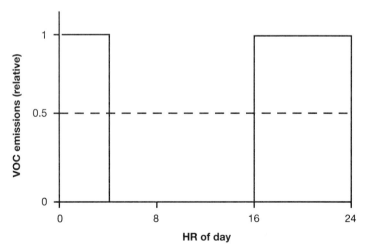

FIGURE 5.2 Illustrative patterns of VOC emissions from a manufacturing facility. Constant emissions have a significant impact on photochemical smog generation, while emissions only in the 4 PM to 4 am period do not.

anticipated effects of the emission of molecules, such as carbon dioxide, that are capable of absorbing heat radiation are expected to manifest themselves only gradually, over a time period of at least half a century. Most people find it difficult to relate to a time period longer than a normal adult life span and on such a basis to take actions that may be inconvenient or costly to them in the short run, so analytical approaches that emphasize temporal differences tend to be discounted.

A contemporary example that illustrates the temporal scale conundrum is an effort by Dutch researchers to perform an LCA on the renovation and maintenance of dikes. There are obvious short-term considerations here, such as the types of maintenance equipment and materials to be used, for example. There are also long-term considerations, such as the impact of dikes on ecosystems over the next century or two. It has been proposed that a way to proceed may be to conduct a *cascade LCA*, in which separate assessments are performed for different epochs. The idea has intellectual promise, but it cannot be evaluated until it is reduced to practice.

5.4.3 Spatial Scales

Environmental impacts occur on all spatial scales from local to global. Consider first a global impact such as ozone depletion. This effect does not necessarily occur near a region emitting ozone-depleting species, but rather at a distant location and in combination with the ozone-depleting effects of other emissions of the same type. Consequently, it is legitimate to relate a portion of the global effect to the ozone-depleting species emitted in a particular region.

Consider next a regional impact such as acid deposition. If that deposition occurs in a poorly buffered region such as western Sweden, the effect per molecule of SO_2 emitted will be much greater than the same deposition onto the well-buffered soils of Germany. Alternatively, deposition in identical amounts to identical soils can produce different effects if other depositions are occurring simultaneously.

In both of the preceding cases, the relationship between loading and effect is a function of the environmental and industrial properties of the location where the loading occurs. In general, the uncertainty produced by the assumption of a single loading-effect relationship for the planet rather than a location-specific one is increasingly severe as the distance scale becomes smaller and smaller. For example, the emission of potentially smog-forming hydrocarbons in Calgary is of minor importance, as smog is not a common problem there. The same emission in Los Angeles, however, will exacerbate an already difficult situation and is therefore highly important. Similarly, emissions leading to acid precipitation on well-buffered soil are considerably less serious than identical emissions upwind on poorly buffered soil. Thus, no unique set of environmental indices for emissions or other actions of concern (such as those of the EPS system of Chapter 4) can be utilized in LCAs unless all locale-dependent effects are ignored.

Even if locale-dependent assessments are contemplated, it must be recognized that effects in a locale may vary over time. Emissions occurring during the smog season are more important than those outside the season, for example, and the closing of a major industrial facility or the installation of air pollution controls on a large emissions source nearby may transform relatively unimportant environmental effects into dominant ones. Meteorology and other local environmental factors may even play a role: The discharge of pollutants into a fast-flowing river may be of much less concern than the same discharge into the same river at a time when water flow is low.

5.4.4 Valuation

Achieving worldwide agreement on valuation seems unlikely. Attempts have been made, as discussed in Chapter 4. Without making a judgment on the appropriateness of specific rankings, the methodologies exemplified by the Volvo/Swedish Environmental Research Institute/Federation of Swedish Industries collaboration and by the Netherlands NSAEL method seem to be promising approaches to implementing quantitative impact prioritization, especially if an economic impact analysis along the lines of the Netherlands VNCI approach can be simultaneously incorporated. However, serious limitations remain. Both approaches treat only a subset of environmental concerns, EPS with its five "safeguard subjects" and NSAEL with its five "environmental aspects." Even if all relevant concerns were included, the techniques for arriving at the EPS environmental indices and the NSAEL levels are not generally accepted as scientifically sound. This situation reflects, of course, not only the enormous complexity of environmental science, but also gradual and discontinuous changes in new knowledge of natural systems. Finally, neither the EPS nor NSAEL method attempts the crucial but highly contentious step of valuation.

From the standpoint of both regulated and regulating sectors, it is important that any LCA impact assessment system be regarded as unbiased. For this purpose, it is desirable that a competent third party perform the actual assignment of weighting factors, rather than an individual corporation as was done in Sweden. With the increasing importance of multinational firms, it would be even more helpful if such a system were generated by an international organization such as a committee of the International Geosphere-Biosphere Program (IGBP). Regardless of the organization performing the prioritization, provisions for regular updates are crucial in view of the evolving nature of environmental science.

It is perhaps not so contentious to define the important topics for valuation; the examples of the international treaties on ozone protection and global climate change attest to that. It is another matter altogether to assign numerical weighting factors to each one. A rise in sea level is obviously of more concern to citizens of low-lying Bangladesh than those of mountainous Tibet, and a decrease in visibility more important to those within everyday sight of Mt. Fuji than to residents of central Tokyo. And valuation must be done on a global basis if at all, because of the globalization of product manufacture, sales, and service.

In the opinion of the author, we can hope on a global basis to do no better at valuation, at least in the near term, than to group potential environmental impacts into classes such as "very important," "important," "of lesser importance," and "unimportant." If agreement is achieved on taking a specific set of impacts and "binning" them by that qualitative approach, one can then proceed to incorporate the results into useful, if perhaps not numerically detailed, schemes for assessing the impacts of products, processes, facilities, and services over their lifetimes.

76 Chapter 5 Evaluating the LCA Process

5.5 PONDERING LCA CASE STUDIES

5.5.1 Outdoor Clothing

A life-cycle assessment of environmental influences due to clothing manufacture and distribution was carried out for Patagonia, Inc. by UCLA. A focus of the study was the energy used at two life-cycle stages, manufacture and transportation. During the manufacturing stage, energy is consumed in production and processing of the fiber, and in dying, weaving, and assembling. Transportation can be by any of several modes, but is usually by truck or air. Figure 5.3 shows the balance between the energy consumption attributable to the two life stages; it differs greatly depending on whether truck transport or air transport is used. The study has spawned increased efforts by Patagonia to avoid air shipment where possible.

5.5.2 Polyester Blouses

Although many products consume no energy and produce no residues when in use, some cause the bulk of their environmental impact after manufacture and before disposal. A common example is clothing, which undergoes many washing cycles during its lifetime, thereby consuming energy in heating water for washing and air for drying as well as requiring resources for the production and disposal of the detergents. An assessment of these environmental interactions for the case of a women's knit polyester blouse has been prepared by Franklin Associates for the American Fiber Manufacturers Association.

Figure 5.4a shows the total energy requirements per million wearings (this turned out to be an appropriate functional unit) for the blouses. The energy used during manufacture (Fig. 5.4b) is divided as would be anticipated, resin production and

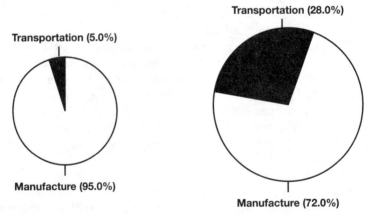

FIGURE 5.3 Relative energy consumption during manufacturing and transportation for a Patagonia, Inc. garment. (left) Truck transportation; (right) Air freight transportation. The total energy consumed is indicated by the area of the circle. (Adapted from L. Hopkins, D.T. Allen, and M. Brown, Quantifying and reducing environmental impacts from transportation of a manufactured garment, *Pollution Prevention Review, 4* (1994) 491-500.)

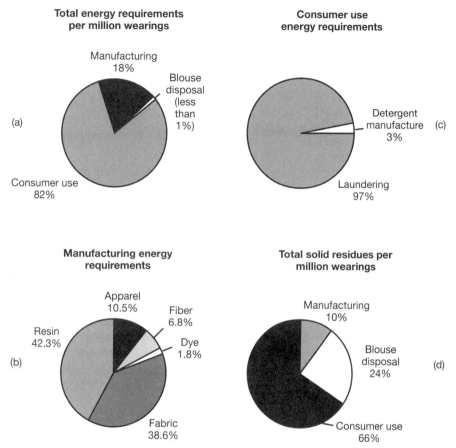

FIGURE 5.4 Data from an inventory assessment analysis of women's polyester blouses. (a) Allocation of total energy requirements per million wearings. (b) Allocation of manufacturing energy requirements. (c) Allocation of consumer use energy requirements. (d) Allocation of total solid residues per million wearings. (Reproduced with permission from American Fiber Manufacturers Association, *Resource and Environmental Profile Analysis of a Manufactured Apparel Product: Woman's Knit Polyester Blouse*. Copyright 1993 by American Fiber Manufacturers Association.)

fabric production requiring roughly equivalent amounts and other activities being much less important. Manufacturing energy is less than a fourth of that required for consumer use. The latter is allocated in Figure 5.4c, and is seen to be almost entirely due to laundering operations. In fact, if energy alone were a consideration, total environmental impacts would be minimized if blouses were replaced after every fourth wearing rather than being repeatedly laundered. Solid residues also occur predominantly during the customer use phase (Fig. 5.4d); these are the municipal sludge attributable to the washing operation and the ash related to off-site energy generation.

What changes in customer use patterns could improve the environmental responsibility of the blouse design? The study found that use of a cold wash cycle, thus eliminating the need for heating water, reduced overall laundering energy consumption by 60 percent. A less extreme alternative is lowering the water temperature but retaining a warm wash; a 10 degree temperature reduction reduced laundering energy use by 14 percent. Line drying was also a very beneficial activity, reducing overall energy consumption by 31 percent.

What do these results say to the product designer? An obvious answer is that she or he should try to develop blouses that can be cleaned effectively in cold water, perhaps the easiest change that may influence consumer behavior. A second potential change is to modify the product so that it line-dries quickly or dries mechanically in a shorter period of time while retaining an attractive appearance. Probably less effective, but still worth doing, is encouraging in the product labeling the use of lower temperature water if a warm-water cycle is used, and of air drying.

5.5.3 Grocery Sacks

The Council for Solid Waste Solutions has conducted an LCA on the relative merits of polyethylene and paper grocery sacks. The life stages that were the focus of the study were manufacturing and disposal and the functional unit was one sack's worth of carrying capacity. Two crucial assumptions had a major influence on the results: the number of polyethylene sacks used instead of a single paper sack, and the rate of recycling of the sacks. In the former case, the analyses were performed with polyethylene to paper ratios of 1.5:1 and 2:1, thought to encompass the range of common practice. To assess the influence of recycling rate, all studies were carried out for rates between 0 and 100 percent.

The results of the study are shown in Figure 5.5. In every topic evaluated—energy use, solid waste produced, atmospheric emissions, and waterborne wastes—the polyethylene sacks were superior to the paper sacks. Only in the specific subcategory of volatile hydrocarbon emissions (not shown) were the paper sacks better. Biodegradability was not a discriminant, because the available data indicate that in a landfill that is top-covered each day little degradation of either material is evident.

The analysis did not attempt to evaluate the relative merits of the individual environmental stressors with respect to each other.

5.5.4 Hot Drink Machines

The University of Amsterdam has conducted a partial LCA on the manufacture and use of machines that dispense hot drinks. The boundaries of the study were quite broad, including the extraction and production of materials related to the hardware, extraction and production of the hot drink ingredients, and construction, use, and disposal of the machines. Among the results were the following:

- The largest amount of waste occurs in the in-use phase of the machine (coffee, sugar, cups, etc.).
- The largest amount of energy use occurs while the machine is on stand-by as it keeps water hot for instant use.

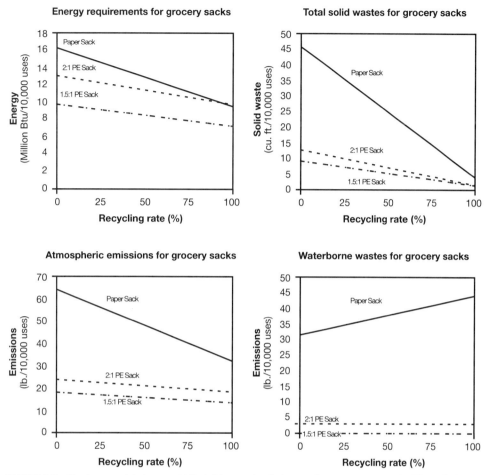

FIGURE 5.5 Environmental stresses attributable to polyethylene and paper grocery sacks. The units in each case are per 10,000 uses, and the analyses are performed for the cases in which 2.0 and 1.5 polyethylene sacks are used in place of each paper sack. (Adapted with permission from The Council for Solid Waste Solutions, Copyright 1993.)

- The use of a better insulated water heater would minimize energy consumption with little design change.
- The addition of a "no cup" button for those using their own mugs would decrease in-use waste.

In this study, the LCA analysis was never extended past the inventory phase, though even that level was obviously sufficient to produce useful results.

5.5.5 Food Products

The Swedish Waste Research Council has conducted a review of the life-cycle assessments of various food products: milk, fruit juices, bakery products, margarine, goulash, and meats. The individual reports examined waste generation, production requirements, packaging, transport, and consumption. Although specific to the grocery business, many of the results suggest product design and assessment principles that are applicable to products other than food.

One characteristic of packaged foods is that a significant percentage of the food products will never be consumed. For example, some 15 percent of bread is wasted as a result of becoming stale prior to sale. Similar amounts of fresh produce are lost to decay. For liquid products, fractions of 1–20 percent (depending largely on viscosity) are retained on the walls of the containers in which the products are packaged. In all these cases, improved packaging is likely to help, although the environmental impacts of manufacturing the packaging itself must be considered.

The environmental impacts of product distribution are often very significant for food products. In the case of produce, for example, the impacts are relatively small for locally produced fruits and vegetables. If fruits and vegetables are imported from other continents in order to render the supply relatively independent of the time of year, however, the chargeable energy costs become very high.

The provisioning of appropriate sizes of food products is very important. If the quantity is larger than is desired by the customer, the remainder will often be discarded, especially in the case of individually packaged "convenience foods." The wasted energy is sufficiently high that packaging of most food products in many more sizes than are now available can readily be justified on environmental grounds.

In general, the DFE assessment of food products is in a rudimentary stage, with the inventory step of the LCA being incompletely performed and the impact analysis being ignored in all of these studies.

5.5.6 Computer Workstations

The Microelectronics and Computer Technology Corporation, an industry consortium, has conducted what is termed a life-cycle assessment of a computer workstation. The workstation components were defined as follows:

- One 6-inch (diameter) silicon wafer of components
- 220 integrated circuits
- 517 square inches of single and multi-layer printed wiring board
- One monitor with a diagonal length of 20 inches

Separate analyses were conducted for semiconductor device fabrication, semiconductor packaging, printed wiring board manufacture, display manufacture, and systems assembly. Figure 5.6a is the materials and energy flow diagram for display manufacturing (one of several similar input/output diagrams), and Figure 5.6b shows a flow diagram for the entire system.

Some of the results of the assessment are shown in Figure 5.7. At the top is the energy consumption diagram, which demonstrates that the customer use phase completely dominates the energy requirements of the workstation. It was analyses like this that led the computer industry and the U.S. Environmental Protection Agency to

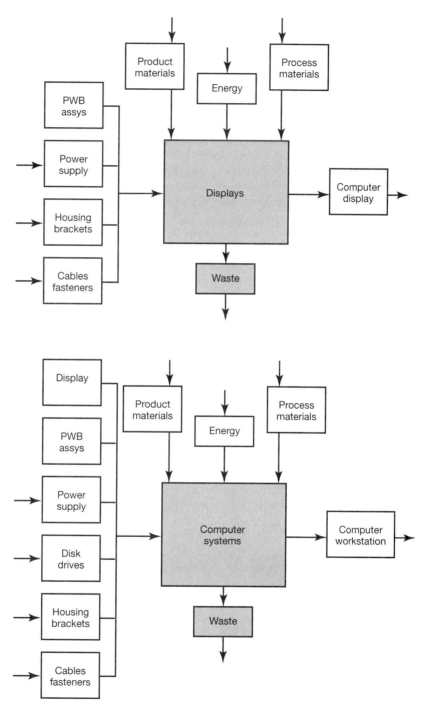

FIGURE 5.6 Material and energy flow diagrams for a computer workstation. (Top) the 20-inch monitor; (bottom) the workstation system. (Reproduced with permission from Microelectronics and Computer Technology Corporation, Copyright 1994.)

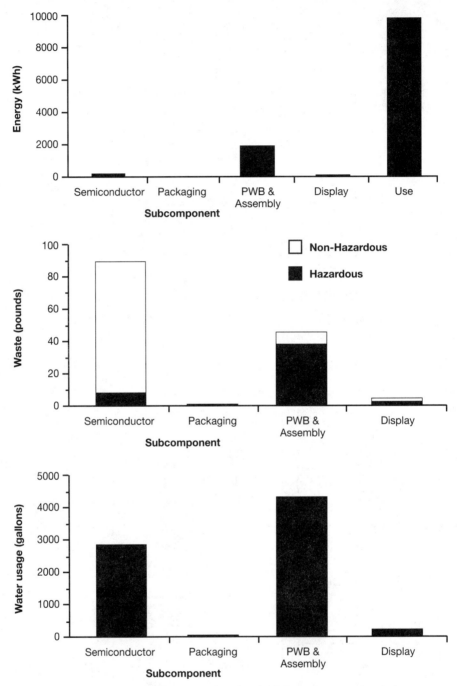

FIGURE 5.7 LCA results for a computer workstation. (top) Energy consumption during production and use; (middle) waste generation during manufacture; (bottom) water usage during workstation manufacture. (Reproduced with permission from Microelectronics and Computer Technology Corporation, Copyright 1994.)

establish the Energy Star program, which promotes the development of computers that revert to low-energy "sleep mode" when not in active use.

The waste generation diagram for the workstation manufacturing phase is given in the center of Figure 5.7. The total waste generation is about 150 pounds per workstation, about 50 pounds of which is classified as "hazardous," that is, waste that must be recycled or the hazardous constituents recovered. The waste at the semiconductor stage is largely scrap from the wafer processing, while the waste from the printed wiring board manufacture contains copper and other metals in solution. Little effort need be expended on reducing waste from the semiconductor packaging or display manufacturing stages, as the waste generation is small.

The bottom of Figure 5.7 shows the water use during the manufacture of the components. One computer workstation is responsible for some 7,000 gallons of wastewater, generally in the various cleaning and washing stages. This wastewater must be treated to remove any harmful impurities that are picked up by the water during the manufacturing process. In addition, most of the water used at the semiconductor manufacturing stage is deionized, an energy-intensive process. High water usage is an area where substantial potential exists for improvement.

The findings and recommendations from the study mention most of the items discussed above solely from the perspective of an energy and materials inventory. No attempt is made to perform an impact analysis. As is said in the report, "Since an absolute measure of environmental impact does not exist, one environmental option cannot absolutely outweigh another."

5.5.7 Disposable and Reusable Diapers

The comparative environmental merits of disposable and reusable (i.e., cloth) diapers have been assessed not by one organization, but by more than a dozen. Each study looked at materials and energy flows at the different life stages: raw materials extraction, manufacture, product delivery, use, and disposal.

The items included in the inventory analysis are relatively straightforward. The raw materials stage involves water, pesticides, and energy needed to raise and recover the cotton for the reusable diaper, or the energy and associated solid waste and air pollution involved in extracting and processing the wood fiber and petroleum needed for the disposable diaper. The manufacturing stage involves modest amounts of energy and processing chemicals in each case, and stresses on the environment during product delivery are essentially identical. After the diaper is transferred from the manufacturer or dealer to the customer, the disposable diaper adds substantial amounts of once-used material and human wastes to landfills, while reusable diapers consume large amounts of energy and detergents in laundering.

The approaches of this spectrum of studies turn out to be significantly different, and depend to some extent on the assumptions made: the frequency of use of diapers, the number of rewashings possible, the price of energy, and so forth. Results have been expressed as the amount of material used in each life stage, the amount of energy, or some other convenient unit. Partly as a consequence of differing assumptions and different assessment boundaries, neither type of diaper is unequivocally recommended, but customers living in areas where landfill space is scarce and expensive are advised to use cloth diapers, while those living where water is scarce are advised to use disposables. This latter advice amounts to a weak attempt to use locale-dependent valuation.

5.5.8 Retail Sale of Building Materials

The University of Amsterdam has conducted a partial LCA on the products and operations of a chain of retail building materials stores. Rather than assess all product lines, the initial LCA was performed on doors made from tropical hardwood and from European Douglas fir. The inventory analysis is relatively simple, consisting of receipt of the wood, cutting and assembling of the door, and painting of the wood with a primer. The recommendations include reductions in wood waste through improved parts cutting, reducing the emissions from the painting operations, and choosing to make the door from European timber in preference to using tropical hardwoods. The first two of these recommendations are solely inventory analysis-related, the last reflects a value judgment that tropical hardwoods are a scarce resource whose use should be minimized, regardless of the possible effects on European timber.

5.5.9 Lithographic Printing Operations

The Battelle Corporation has conducted an LCA on the lithographic printing process. It is of interest that the focus was not on a specific product, but on the process by which a product is produced. Considerable attention was given to identifying the chemicals used at each stage, as well as the quantities consumed.

This study differs from most of the others discussed here in that an impact analysis was performed; that is, the environmental impacts that would result from lithographic printing activities were assessed. The impact ratings are completely arbitrary: some relate to regulatory limits, some to industry norms, some to the use of specific chemicals. In addition, no attempt was made at prioritization, so, for example, smog-forming potential was weighted equally with global-warming potential. The result is a mixing of temporal scales, spatial scales, and impact magnitude, with no attempt at justifying the impact assessment approach.

5.5.10 Case Study Summary

While the case studies presented above and those discussed in previous chapters do not represent the totality of all LCAs conducted to date, they are an illustrative sample. The picture that emerges is one in which many of the studies restrict themselves to an inventory analysis, and then seek to identify at which life stage the dominant energy use occurs, or the dominant emissions, or the dominant solid waste. The recommendations then concentrate on those environmental concerns and those life stages. There are several implications here: that "less is better," regardless of less of what, that all environmental concerns are of equal importance, and that environmental concerns not treated are of no importance whatever.

In a few cases where an attempt is made to perform an impact analysis, the analysis is either not prioritized, or is prioritized by a technique not generally accepted by the DFE community.

5.6 ASSETS AND LIABILITIES OF THE LCA PROCESS

The great asset of the LCA concept is that it appears inherently to be the ideal way to quantitatively assess the range of environmental impacts attributable to a specific product. As seen in this chapter, however, there prove in practice to be many limitations. The drawing of assessment boundaries is difficult, the specification of the functional unit is not obvious, and it is difficult to recommend a consistent approach to either. Data collection and analysis are also limitations to accuracy and completeness. As a result, inventory analyses by different assessment teams can produce different, though perhaps equally defendable, results.

Still more intractable are the problems of impact analysis, especially variations in temporal scale, spatial scale, and locale. Perhaps most difficult of all are the questions of valuation, in which the absolute assignment of value to different environmental impacts is thwarted by differences in societal structure and preference. These constraints lead at least a few practitioners to say that LCAs can only study burdens placed on the environment and not environmental impacts. Because quantifying the latter is the purported reason for doing LCAs, however, a retreat to burdens is, in a sense, a retreat from the desired quantitative approach. Finally, as T.K. Boguski of Franklin Associates points out, no matter how sophisticated a quantitative analysis may be, if it has a subjective basis or uses subjective data, it gives subjective results.

Some of these difficulties, though by no means all, are less serious if LCAs apply not to a single product or a single corporation but to a large geographical or political unit. The LCA for a region, country, or continent, for example, can more reliably incorporate average values for energy cost and use summed resource flows rather than worry about attributing specific portions of the flows to specific products. Questions of temporal scale and lack of data can then be addressed more directly, and the results of a policy-centered LCA used as a guide to governmental purchases, agreements, and regulations.

Is, then, the concept of a product-level comprehensive LCA, with its scoping, inventory, impact, and interpretation phases, infeasible, at least as a routine tool? In the LCA's present form, the answer is "probably." Nonetheless, those who have performed almost any of the types of LCAs mentioned above have found them to benefit both the product being assessed and the environment that is affected. That fact suggests that a less doctrinal and simpler version of an LCA might have substantial utility, whether or not it meets all the lofty goals of the ultimate LCA. This simpler approach, the streamlined LCA, is the subject of the remainder of this book.

FURTHER READING

LCA Techniques Overviews: L. G. Lindfors, et al., *LCA-NORDIC Technical Reports No. 1-9* and *LCA-NORDIC Technical Report No. 10* and *Special Reports No. 1-2*, Rpts. Tema Nord 1995:502 and 1995:503 (Copenhagen: Nordic Council of Ministers, 1995); J.W. Owens and S. Rhodes, *Discussion Paper on LCA Impact Assessment Category Feasibility*, Rpt. ISO/TC 207/SC 5/WG 4N30 (Geneva: International Organization for Standardization, 1995); J.W. Owens, LCA impact assessment categories: Technical feasibility and accuracy, *International Journal of Life Cycle Assessment, 1* (1996) 151–158; J. W. Owens, Life-cycle assessment: Constraints on moving from inventory to impact assessment, *Journal of Industrial Ecology, 1* (1997) 37–49.

Manufacture and Transportation of Garments LCA: L. Hopkins, D.T. Allen, and M. Brown, Quantifying and reducing environmental impacts from transportation of a manufactured garment, *Pollution Prevention Review, 4* (1994) 491–500.

Polyester Blouses LCA: American Fiber Manufacturers Association, *Resource and Environmental Profile Analysis of a Manufactured Apparel Product: Woman's Knit Polyester Blouse* (Washington, D.C., 1993).

Grocery Sacks LCA: The Council for Solid Waste Solutions, *Resource and Environmental Profile Analysis of Polyethylene and Unbleached Paper Grocery Sacks* (Washington, D.C., 1993).

Hot Drink Machines LCA: R. van Berkel, et al., in *Ecodesign: Eight Examples of Environment Driven Product Development*, H. Riele and A. Zweers, Eds. (The Hague: SDU Publishers, 1994) pp. 69–88.

Food Products LCA: K. Anderson, T. Ohlsson, and P. Olsson, *Life Cycle Assessment of Food Products and Production Systems. II. LCA and Foods*, AFR Report 26 (Stockholm: Swedish Waste Research Council, 1993).

Computer Workstations LCA: *Life Cycle Assessment of a Computer Workstation*, Rpt. HVE-059-94 (Austin, TX: Microelectronics and Computer Technology Corporation, 1994).

Disposable and Reusable Diapers LCA: C. Lehrburger, *Diapers in the Waste Stream: A Review of Waste Management and Public Policy Issues* (Sheffield, MA, 1989); *Disposable Versus Reusable Diapers: Health, Environmental, and Economic Comparisons* (Cambridge, MA: Arthur D. Little, Inc., 1990).

Retail Sale of Building Materials LCA: R. van Berkel, et al., *Environmental Improvements in Retail Trade: Interim Report of the PRIMA Initiative* (IVAM Environmental Research, University of Amsterdam, 1994).

Lithographic Printing Operations LCA: D.A. Tolle, B.W. Vigon, J. Becker, and M.A. Salem, *Development of a Pollution Prevention Factors Methodology Based on Life-Cycle Assessment: Lithographic Printing Case Study*, Rpt. EPA/600/R-94/157 (Cincinnati, OH: U.S. Environmental Protection Agency, 1994).

EXERCISES

5.1. In Section 5.3, the problem of deciding whether or not to use modern energy, emissions, and impact data to assess old facilities was presented. Two points of view that could be taken on the problem are: (1) Old data should be used because those data apply to the situation when the facility was built, or (2) current data should be used because that produces an equivalent assessment. Adopt one of these two views, defend your choice, and comment on the hydroelectric/gas assessment approaches from that perspective.

5.2. Refer to Figure 5.1a. Assume that you are attempting to evaluate the response produced by a loading slightly greater than L_r and you suspect the relationship may have a threshold and may be nonlinear. If the true relationship is that of Figure 5.1b, will you overestimate or underestimate the response by using the linear relationship? Repeat for Figure 5.1c and Figure 5.1d. Repeat the entire exercise if the loading is slightly less than L_r.

5.3. In the studies for which results are pictured in Figures 5.3–5.6, what functional units were used? Explain why these were or were not good choices.

5.4. Assume that a valuation factor of 1.0 (highest value), 0.5 (moderate value), or 0.0 (no value) was to be assigned to four environmental concerns: ozone depletion over Antarctica, southern Argentina, and southern Australia; a large oil spill in Hamburg harbor; a 20 percent loss in global biodiversity; radioactive emissions from the Chernobyl nuclear power plant. What factors should you assign if you are a resident of (a) Hamburg, Germany; (b) Sydney, Australia; (c) Beijing, China; (d) Minneapolis, USA?

CHAPTER 6: The Streamlined LCA Philosophy

6.1 THE ASSESSMENT CONTINUUM

If no limitations to time, expense, data availability, and analytical approach existed, a comprehensive LCA as described in Chapters 2–4 would provide the ideal advice for improving environmental performance. In practice, however, these limitations are always present. As a consequence, although very extensive LCAs have been performed, a complete, quantitative LCA has never been accomplished, nor is it ever likely to be. There are many compromises of necessity, among which have often been the use of averages rather than specified local values for energy costs, landfill rates, and the like, the omission of analysis of catalysts, additives, and other small (but potentially significant) amounts of material, neglect of capital equipment such as chemical processing hardware, and the failure to include materials flows and impacts related to supplier operations. As a consequence, detailed LCAs cannot be regarded as providing rigorous quantitative results, but rather as providing a framework upon which more efficient and useful methods of assessments can be developed.

The question of data availability as it relates to product design and manufacture deserves added discussion. Experts agree that roughly 80 percent of the environmental costs of a product are determined at the design stage, and that modifications at later stages of product development will have only very modest effects. The ideal time, then, to conduct an LCA analysis is early in the design phase. At that point, however, the characteristics of the product tend to be quite fluid: materials may not have been selected, no manufacturing facility may have been built, no packaging approach may have been determined, and so on. Hence, there is often no possible way to complete a quantitative LCA at the precise time when one would be most useful.

Techniques that purposely adopt some sort of simplified approach to life-cycle assessment, *streamlined life-cycle assessments,* form part of a continuum of assessment effort, with the degree of detail and expense generally decreasing as one moves from the left extreme toward the right, as shown in Figure 6.1. The region termed *extensive LCA* is that of detailed, quantitative LCAs such as the IVL/Volvo front-end analysis discussed in Chapter 4. The *scoping* or *ecoscreening* regions are those that are purposely sketchy, done to ensure that no truly disastrous design choices have been made or to determine whether additional assessment is needed. Somewhere within the SLCA region is the ideal point: The assessment is complete and rigorous enough to be a definite guide to industry and an aid to the environment, yet not so detailed as to be difficult or impossible to perform.

88   Chapter 6   The Streamlined LCA Philosophy

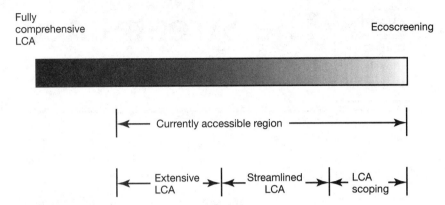

FIGURE 6.1   The LCA/SLCA continuum.

6.2   PRESERVING PERSPECTIVE

If streamlining is to be a universal characteristic of LCAs, how can one tell if a simplification has not streamlined away an assessment's legitimacy? In a survey of ways in which practitioners from academia, government, industry, and consulting firms are attempting to abridge the LCA, Keith Weitz of North Carolina's Research Triangle Institute and his coworkers identified the nine approaches that are discussed below:

- *Screen the product with an inviolates list.* This approach treats some activities or choices as so obviously incorrect from an environmental standpoint that no design or plan to which they apply should be allowed to go forward. Examples of inviolates are the use of mercury switches in a product or of CFCs in manufacturing. While an inviolates list is useful as an assessment tool, limiting an assessment to the use of such a list obviously has the potential to overlook many life stages and environmental stressors.
- *Limit or eliminate components or processes deemed to be of minor importance.* Some studies use weight limits or other factors to minimize the amount of detail required.
- *Limit or eliminate life-cycle stages.* Some studies limit the LCA to practices occurring within an industrial facility. This "gate-to-gate" approach amounts to a version of "pollution prevention." While meritorious, it clearly does not satisfy the criterion of treating the entire life cycle. A second common approach is to limit or eliminate only upstream stages (resource extraction, for example). This approach is potentially broader than gate-to-gate, especially if the evaluation of upstream stages is limited rather than eliminated. A third alternative, potentially useful for materials suppliers having little control over the use of their products, is to limit the LCA to life stages 1-3, i.e., "cradle to warehouse".
- *Include only selected environmental impacts.* Some studies limit the LCA to impacts of highest perceived importance or those that can be readily quantified. Such choices may often be responsive to public pressure rather than to environmental science, and to be anthropocentric rather than balanced.

- *Include only selected inventory parameters.* This is a variation of the preceding approach, because if only selected impacts are of interest, only the inventory data needed to evaluate those impacts will be gathered.
- *Limit consideration to constituents above threshold weight or volume values.* An assessment may be limited only to major constituents or modules. This limitation overlooks small but potent constituents (it would fail as a tool for an SLCA of medical radioisotope equipment, for example), but may sometimes be justifiable from the standpoint of efficiency and tractability. It obviously applies only to quantitative assessments.
- *Limit or eliminate impact analysis.* Impact analysis is a major component of LCA, and eliminating it clearly abridges the process. The result is that the overall assessment can rely only on a "less is better" philosophy. While pursuing such an approach will probably result in some useful actions, the approach provides absolutely no connection between the knowledge base of environmental science and the recommendations made by the abridged LCA.
- *Use qualitative rather than quantitative information.* Quantitative data are often difficult to acquire or may not even exist. Conversely, qualitative data can be sufficient to reveal the potential for environmental impacts at different life stages. However, the qualitative approach can make it difficult to compare one product with another or with a new design if the ratings are quite similar.
- *Use surrogate data.* It is sometimes possible to use data on a similar material, module, or process when the specific data desired for an assessment are not available. The use of surrogate data is often contentious, and has many of the same limits in usefulness as qualitative data.

One additional approach not mentioned by Weitz and colleagues is

- *Eliminate interpretations or recommendations.* In some studies, inventory and impact results are provided in detailed reports, with the recipient left to devise actions that should be taken in response to the report. If an SLCA is to be useful, however, specific recommendations should be provided by the assessment team, and a method for implementing those recommendations developed.

The result of this survey of approaches indicates that a valid SLCA should have the following characteristics:

- All relevant life-cycle stages should be evaluated in some manner.
- All relevant environmental stressors should be evaluated in some manner.
- The SLCA should include LCA's four elements: goal and scoping, inventory analysis, impact analysis, and interpretive analysis. These need not necessarily be approached in a quantitative manner.

6.3 ALTERNATIVE SLCA APPROACHES

Several different approaches have been put forward by individual corporations, consulting firms, and professional associations as suitable techniques for SLCA. Their intent is to preserve the LCA concept and rigor sufficiently to inspire confidence in the results, while at the same time meeting the scientific and logistical constraints that are inevitably present. The degree to which these goals are met vary widely.

Many SLCA approaches have gravitated toward a matrix, one dimension of which is life-cycle stages and the other of which is a list of environmental impacts, potential employee health concerns, or other relevant parameters. Such a matrix has the potential for satisfying the first two characteristics of a valid SLCA listed above. Whether the third characteristic is satisfied depends on the degree to which the three LCA stages are addressed. Inventory analysis is almost always covered in SLCAs, at least qualitatively. Impact analysis is often not covered, or poorly incorporated from an environmental science standpoint. Interpretation is almost always inferred, but improvements may not occur unless the overall approach involves a structured technique for making and implementing recommendations.

A number of these different SLCA approaches are described below. The intent is not that this be a comprehensive list, but that it illustrate some of the favorable attributes as well as the deficiencies of techniques presently used for SLCAs.

6.3.1 THE MIGROS CONCEPT

Migros is the largest chain of grocery stores in Switzerland. The organization conducts SLCAs on the packages used for products sold in its stores. Packaging is the only life stage treated, and the environmental stresses addressed are limited to four: energy consumption, air burden, water burden, and fixed waste. The technique is used only for comparison of alternate packaging approaches, not as an absolute measure. An impact-valuing scheme is employed, but it is proprietary and not obviously related to environmental science. Hence, the Migros effort limits both life-cycle stages and environmental stress parameters.

6.3.2 UNIVERSITY OF BRITISH COLUMBIA'S AND IBM CORPORATION'S SLCA APPROACH

The University of British Columbia was among the first to conduct detailed inventory analysis, the subject being the comparison of the resource use, energy use, and emissions chargeable to hot-drink cups of paper and plastic, which was discussed in Chapter 3. A similar approach has been taken by the IBM Corporation, which has studied the resource and energy flows involved in alternative designs of personal computers. In neither of these approaches was any attempt made to determine the relative environmental effects produced by the different products. Accordingly, these approaches can be classified among those that eliminate impact analysis, although they do an excellent job of inventory analysis.

6.3.3 THE DOW CHEMICAL COMPANY MATRIX

The Dow approach defines a matrix of eight life stages and 12 impact parameters, as shown in Figure 6.2. The impacts are not all strictly environmental, and include safety,

Section 6.3 Alternative SLCA Approaches 91

Environmental Dimensions	RME	RMP	MFG	DST	CNV	NDU	DSP	RCY
Safety: Fire, Explosion								
Human Health								
Residual Substances								
Stratospheric Ozone Depletion								
Air Quality								
Climate Change								
Natural Resource Depletion								
Soil Contamination								
Waste Accumulation								
Water Contamination								
Public Perception Gap								
Competition								

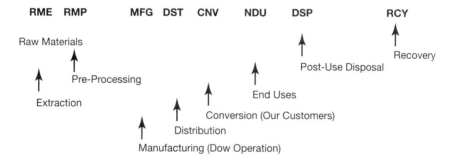

RATING SYSTEM

Hazard or Effect	Exposure (Volume, Frequency)		
	HIGH	MEDIUM	LOW
HIGH	-9	-9	-9
MEDIUM	-9	-3	-1
LOW	-3	-1	

9	Proven Implemented Solution
3	Project Initiated Resources Allocated
1	Project Identified

Vulnerabilities   *Opportunities*

FIGURE 6.2 The Dow Chemical Company SLCA matrix (courtesy of S.D. Noesen).

public perception, and competitive advantage. Scoring is done in two steps. First, the user determines whether the matrix element represents a vulnerability or an opportunity. Next, the scoring of the element is done on a three-level (1, 3, 9 or −1, −3, −9) scale.

In principle, the Dow matrix includes all life-cycle stages and environmentally related concerns. It is a qualitative system.

6.3.4 THE MONSANTO MATRIX

As does Dow, Monsanto incorporates health and social impacts into its assessment matrix, shown in Figure 6.3. This grouping of topics occurs largely because these functions, as well as more traditional concerns such as emissions and waste generation, are generally addressed by the same staff organization within a firm. The matrix explicitly separates inventory issues from impact issues. It also explicitly includes business impacts, which amounts to incorporating LCA's interpretation stage, though not as comprehensively as the Pareto approach presented in Chapter 4.

Guidelines for matrix element evaluation provide the means by which the concerns and knowledge of environmental science, ergonomics, and other specialties are incorporated. The set of guidelines is specific to Monsanto and is inevitably a work in progress, but it shares much commonality with those of other manufacturers (as presented, for example, in the appendices of this book).

6.3.5 MOTOROLA'S SLCA APPROACH

The Motorola approach defines a matrix of five life stages and four impact parameters: resource use, energy use, human health, and eco health, as shown in Figure 6.4. The

	Sustainability dimension	Product Life Cycle Stage			
		Raw material	Manufacturing	Product use	Recycle & disposal
Inventory	Resource usage				
	Waste output				■
	Health & quality of life				
Impact	Environmental impacts				
	Social impacts				
	Business impacts				

FIGURE 6.3 The Monsanto Corporation matrix (courtesy of J.P. Mieure).

A life cycle matrix for assessment of a product

Product		Part sourcing	Manufacturing	Transportation	Use	End of life
Sustain-ability	Resource use					
	Energy use					
Human health						
Eco health						

FIGURE 6.4 The Motorola Corporation SLCA matrix (courtesy of W.F. Hoffman III).

concept is that the SLCA process will be performed in three tiers as the product moves from the initial design concept phase (Tier 1) through detailed drawings (Tier 2) to the final product specifications (Tier 3). In Tier 1, four specific questions are asked to determine the evaluation of each matrix element; all can be answered yes or no. The results are used as guidance in the development of design choices. An overall score can be computed by adding all the yes answers, and changes in that score then track overall improvements in the product's environmental characteristics.

Tier 2 assessment in the Motorola scheme is not yet implemented, but the plan is that when the generalized CAD drawings become available, they will be used to provide the basis for semiquantitative scoring of the design. Each matrix element will be scored as a percent of the best possible system, and a 0–100 score for the entire design provided. It is unclear how the "best possible system" will be defined.

Tier 3 is to be performed when all aspects of the design are finalized and (ideally) all quantitative information is available. Thus, a comprehensive LCA can be performed if desired. This phase of the assessment, also not yet implemented, will, of course, be subject to all the caveats surrounding comprehensive LCAs discussed in Chapter 5.

In the ambitious Motorola system, the Tier 1 assessment treats all life stages but only a portion of the potential environmentally related concerns. It is purposely qualitative. The characteristics of the second and third tiers cannot be determined until the techniques are more completely developed.

6.3.6 BATTELLE'S POLLUTION PREVENTION FACTORS APPROACH

Battelle has developed what it calls a *P2 approach* to SLCA that also utilizes a matrix tool. The rows in the matrix are 24 items attempting to cover cradle-to-grave aspects of the life cycle (energy use–raw materials, energy use–product assembly, etc.) and the columns are individual components in the products. An example is shown in Figure 6.5. Each matrix element receives a score based on a five-level scale; the 1, 3, 5, 7, 9 choices placing more emphasis on extremes than on midrange values. The guidance for scoring varies with the product, but often employs either threshold values or emission flux limits.

Summary of Pre-LCA Item Scores for Video Display Shipping/Packaging, including Polypropylene Foam Pad Alternative

		Base Case System						Alt. 1
	Item weight	Crrgtd. case	P-Ureth foam pads	Polyethylene bags	Crrgtd. Kybrd strip	Book wrppr.	Total score	P-propyl foam pad
Component weight (g)		1348.4	374.7	100.0	20.0	3.5	1846.6	143
Component weight (%)		73.0	20.3	5.4	1.1	0.2	100.0	8.9
Energy usage—raw materials	1.6	7	7	7	7	7	7.0	7
Energy usage—product assembly	1.6	9	9	9	9	9	9.0	9
Net water consumption—raw materials	1.6	9	9	9	9	9	9.0	9
Net water consumption—product assembly	1.6	9	9	9	9	9	9.0	9
Airborne emissions	3.1	9	9	9	9	9	9.0	3
Waterborne emissions	3.1	9	9	9	9	9	9.0	*
Solid waste generation rate	7.3	5	7	9	5	7	5.6	9
Industrial waste recycle percentage	7.3	9	1	9	9	5	7.4	9
Recycle content	7.3	7	1	3	7	1	5.6	3
Weight of package	4.2	9	9	9	9	9	9.0	9
Source reduction potential	7.3	5	5	5	5	5	5.0	9
Recyclability potential	7.3	7	1	5	7	1	5.7	3
Product/package reuse	6.3	1	1	1	1	1	1.0	1
Similarity pkg/product material	5.7	5	5	5	5	5	5.0	5
Product disassembly potential	0.0	*	*	*	*	*	0.0	*
Bonding of packaging material	4.7	7	7	9	7	9	7.1	9
Waste to energy value	3.1	5	7	9	5	5	5.6	9
Material toxicity	3.7	5	3	7	5	7	4.7	5
Packaging material volume in landfill	4.2	7	5	9	7	9	6.7	5
Landfill leachate	3.7	7	5	7	7	7	6.6	7
Material persistence	3.7	5	1	1	5	3	4.0	1
Toxic material mobility	3.7	5	7	9	5	7	5.6	9
Inhalation toxicity	3.7	5	3	5	5	7	4.6	5
Incineration ash residue	4.2	6	9	9	6	9	6.8	9
Total	100.0	6.3	4.8	6.6	6.3	5.7	6.0	6.1

FIGURE 6.5 The Battelle Corporation SLCA matrix. (Reproduced with permission from D. Tolle et al., *Proc. 2nd IEEE Int. Symp. on Electronics & Environment,* Rpt. 94CH3386-9, pp. 201-206 [Copyright by Inst. Electrical and Electronics Engrs., 1994].)

After matrix element values are chosen, a weighting factor is assigned to each row "on the basis of Battelle's assessment of the environmental significance (magnitude, geographic scale, location) of an issue," purportedly based on experience and on the published scientific literature. These weighting factors are "further tempered by proportionately re-allocating weight from the materials manufacturing life-cycle stage to product manufacturing and post manufacturing stages." This weighting is an attempt to inject impact analysis into a qualitative SLCA; it may be quite defendable, but, like the Dutch NSAEL approach, will need to be fully explained and then evaluated by the environmental science community.

6.3.7 JACOBS ENGINEERING'S SLCA APPROACH

Jacobs Engineering has developed a matrix tool (Fig. 6.6) utilizing five environmental stresses and seven "risk areas" (global warming, etc.). It has thus far been applied to manufacturing processes, but not to products. The matrix is evaluated for the influence of the process on two spatial scales, local and global. The existing operation is used as a basis from which to evaluate process changes, and the matrix element scores are $+1, 0$, or -1 depending on whether the alternative proposed is better than, equivalent, or poorer than the base case from an environmental standpoint.

The Jacobs approach is the only one of those reviewed here to explicitly evaluate processes and to address the difficult problem of spatial scales. It suffers from a lack of rigorous linking to environmental science in its implementation, and omits many process life stages, such as the construction and eventual disposal of the process equipment.

Impact Analysis Matrix for Aqueous Cleaning Versus Vapor Degreasing

Risk areas / Impacting parameters	Shop level						Global level					
	Material inputs	Energy inputs	Atmospheric emissions	Aqueous wastes	Solid wastes	TOTAL	Material inputs	Energy inputs	Atmospheric emissions	Aqueous wastes	Solid wastes	TOTAL
Global warming		+1	−1			0	0	−1				−1
Ozone depleting resource utilization			−1			−1		−1				−1
Non-renewable resource utilization	−1	+1				0	−1	0				−1
Air quality		+1	−1	+1	−1	0		0	−1	+1	−1	−1
Water quality		+1		+1		+2		0		+1		+1
Land disposal		+1		+1	−1	+1		0		+1	−1	0
Transportation effects	−1	+1			−1	−1	−1	0			−1	−2
Total	−2	+6	−3	+3	−3	+1	−2	0	−3	+3	−3	−5

FIGURE 6.6 The Jacobs Engineering Company SLCA matrix. (Reproduced with permission from M. Callahan, in *Proc. 2nd IEEE Int. Symp. on Electronics & Environment*, Rpt., 94CH3386-9 pp. 201–206 [Copyright by Inst. Electrical and Electronics Engrs., 1994].)

6.4 MATRIX CALCULATIONS

Regardless of how matrix element values are derived, an LCA or SLCA analysis using a matrix-based procedure can be represented mathematically as an exercise in matrix manipulation. To demonstrate, consider the matrix of Table 6.1, which is an SLCA tool devised for ecolabel certification of products in Europe. If the matrix elements $f_{m,n}$ are filled with inventory analysis data, the result is a form of an inventory analysis matrix that can be called **F**. Similarly, matrix elements $s_{m,n}$ can be filled with impact assessment data (a one-time operation except for revisions) to give an impact analysis matrix called **S**.

The (S)LCA assessment for a single critical environmental property n is then given by

$$L_n = \sum_{1}^{m} f_{(m,n)} \times s_{(m,n)} \tag{6.1}$$

In the same fashion, the (S)LCA assessment for a single life stage m is given by

$$L_m = \sum_{1}^{n} f_{(m,n)} \times s_{(m,n)} \tag{6.2}$$

The overall assessment is given by

$$L = \sum_{1}^{m} \sum_{1}^{n} f_{(m,n)} \times s_{(m,n)} \tag{6.3}$$

As with any matrix, some of the **F** matrix elements may contain zeros. This situation will occur in either of two situations: a null inventory value might be listed for such factors as anticipated soil pollution and degradation during the distribution of a product, or where an inventory value may be deemed unimportant. Similarly, zeros will occur in the **S** matrix if no impact is foreseen from a product or process. For wire fencing, for example, no consumption of energy occurs during use no matter how the fencing is designed or from what materials it is constructed.

TABLE 6.1 The European Community Ecolabel Matrix

Critical Property	Product Life-Cycle Stage				
	Materials	Production	Distribution	Utilization	Disposal
Solid waste					
Soil pollution and degradation					
Water contamination					
Air contamination					
Noise					
Consumption of energy					
Consumption of natural resources					
Effects on ecosystems					

This description is, as stated, merely a formalization of the EPS calculation of Table 4.2. There is no reason, however, why the matrix element values must be continuous. Expert systems of various kinds often use data that is quantized: the values may be either binary (as in problem/no problem decision systems) or one of several digits (as in a 1–5 severity ranking system). In either of these cases, the computational approach outlined above is equally appropriate for use. Its important characteristic is that it generates a numerical rating for a design or for alternative designs, thereby providing a target for improvement. As has been said before, "What gets measured gets managed." The result of the SLCA matrix computation fulfills the manager's basic informational requirement, though not with the precision of, for example, financial or production information.

6.5 SLCA ASSETS AND LIABILITIES

As was suggested above, when the LCA concept is streamlined in some way, part of the legitimacy of a comprehensive LCA might be thought to be sacrificed. What is it that is really lost? Conversely, what are the gains? Several of each can be listed and discussed. SLCAs are superior to LCAs in the following ways:

- SLCAs are much more efficient, typically taking several days of effort rather than several months.
- SLCAs are much less costly. They are often capable of being done by existing staff and within existing job requirements.
- Many SLCAs are usable in the early stages of design, when opportunities for change are great but quantitative information is sparse.
- Because of the above attributes, SLCAs are much more likely to be carried out routinely and thus to be applied to a wide variety of products and industrial activities.

SLCAs are inferior to LCAs in the following ways:

- SLCAs have little or no capability to track overall materials flows. Within a corporation, for example, SLCAs on all products might well indicate whether a particular material was used, but not whether its use in a particular product was a significant fraction of total corporate usage.
- SLCAs have minimal capability to compare completely dissimilar approaches to fulfilling a need.
- SLCAs have minimal capability to track improvements over time, e.g., to reliably determine whether a product is environmentally superior to its predecessor.

A caveat on these assets and liabilities harkens back to a statement earlier in this chapter, that a complete LCA has never yet been carried out and probably never will be. That being the case, it seems only reasonable to proceed with the use of SLCAs while recognizing their limitations. The results of SLCAs are often regarded as "approximately correct"; if they come even close to that characterization, carrying out the assessments and implementing their recommendations will be of much value.

FURTHER READING

SLCA Overviews: Weitz, K.A., M. Malkin, and J.N. Baskir, *Streamlining Life-Cycle Assessment: Conference and Workshop Summary Report* (Research Triangle Park, NC: Research Triangle Institute, 1995); Todd, J.A., Streamlining, in *Environmental Life-Cycle Assessment*, M.A. Curran, Ed. (New York: McGraw-Hill, 1996) pp. 4.1–4.17.

The IBM Approach: Brinkley, A., J. Kirby, I.L. Wadehra, P.G. Watson, C. Chaffee, C.N. Cheung, S. Rhodes, M. Wolf, Ecoprofile studies of fabrication methods for IBM computers: Sheet metal computer cover, in *Proc. 2nd IEEE Int. Symp. on Electronics & Environment*, Rpt. 94CH3386-9 (Piscataway, NJ: Inst. Electrical and Electronics Engrs., 1994), pp. 299–306.

The University of British Columbia Approach: Hocking, M., Paper versus polystyrene: A complex choice, *Science, 251* (1991) 504–505, and *252* (1991) 1361–1363.

The Migros Approach: Migros Community Working Group on Environmental Protection, *Migros Environmental Policy* (Zurich: Migros-Genossenschafts-Bund, 1991).

The Dow Chemical Company Approach: Noesen, S., Private communication as reported in Allen, D.A., Applications of life cycle assessment, in *Environmental Life Cycle Assessment*, M.A. Curran, Ed., New York: McGraw-Hill, 1996) pp. 5.1 to 5.18.

The Motorola Approach: Hoffman, W.F. III, A tiered approach to design for environment, in *Proc. Intl. Conf. on Clean Electronics Products and Technology*, Conf. Pub. 416 (London: Institution of Electrical Engineers, 1995) pp. 41–47.

The Battelle Approach: Tolle, D., B. Vigon, M. Salem, J. Becker, K. Salveta, R. Cembrola, Development and assessment of a pre-LCA tool, in *Proc. 2nd IEEE Int. Symp. on Electronics & Environment*, Rpt. 94CH3386-9 (Piscataway, NJ: Inst. Electrical and Electronics Engrs., 1994) pp. 201–206, and Tolle, D.; B. Vigon, J. Becker, and M. Salem; *Development of a Pollution Prevention Factors Methodology Based on Life-Cycle Assessment: Lithographic Printing Case Study*, Rpt. EPA/600/R-94/157 (Cincinnati, OH: U.S. Environmental Protection Agency, 1994).

The Jacobs Engineering Approach: Callahan, M., A life cycle inventory and tradeoff analysis: Vapor degreasing versus aqueous cleaning, in *Proc. 2nd IEEE Int. Symp. on Electronics & Environment*, Rpt. 94CH3386-9 (Piscataway, NJ: Inst. Electrical and Electronics Engrs., 1994) pp. 215–219.

EXERCISES

6.1. Consider the 10 methods of streamlining discussed in Section 6.2. Which of these were used in the Chapter 5 case studies of grocery sacks, food products, and computer workstations?

6.2. Compare and contrast the Dow Chemical matrix in Figure 6.3 with that of Monsanto in Figure 6.4. What are the advantages and disadvantages of each?

6.3. Should an LCA or an SLCA be used in the following situations? Why?

  (a) You are studying the input and output flows of copper in the Canadian economy.

  (b) You are studying the environmental implications of a new design for heavy-lift cranes.

CHAPTER 7

Product Assessment by SLCA Matrix Approaches

7.1 THE MANUFACTURING CHALLENGE

Of all industrial activities, manufacturing is the one that can most easily be environmentally responsible and innovative. Manufacturers are in a unique position. They are not the resource extractor, digging or drilling whatever raw materials have a market; they are not materials processors, forming the powders, crystals, or liquids desired by the component or product assembler; and they are not marketers, making available to customers whatever goods are desired. While those sectors can exercise some influence, they do not have the freedom of the manufacturer, whose sole constraint is to produce a desirable, salable product. In doing so, the manufacturer can choose to make an automobile body from sheet steel, composites, aluminum, or plastic, or a telephone transmission system from coaxial cable, optical fiber, microwave, submarine cable, or satellite systems. She or he can work with "upstream" suppliers to influence the environmental attributes of component design or the choice of materials. Cost, manufacturability, and consumer acceptance limit the choice, of course, but not standard design practice per se. Thus, the designer's role in the manufacturing industry is central.

Although there are substantial commonalities among the manufacturing sectors, there are great diversities as well, as indicated in Table 7.1. An important distinction among the sectors is in the lifetimes of their products. Some are made to function for a decade or more. Others have lives measured in months or weeks. Still others are used only once. A designer obviously must adopt different approaches to these different types of products in terms of durability, materials choice, and recyclability.

In-plant materials and process decisions, traditionally made from the perspective of regulatory constraints and initial cost, can also be made on the basis of environmental preferability. For example, a manufacturing corporation may buy its subsystems from others, but it can exert at least a modest influence on the environmentally related activities of the upstream supplier. Similarly, the manufacturer may choose to perform downstream recycling or at least enter into cooperative arrangements for doing so. In these ways, the manufacturer can examine and influence tradeoffs across the full production-and-use chain, including initial materials choice, product design, manufacturing process specification, in-service impacts, ease of disassembly and reuse, and strategies for recycling and disposal. Even though the manufacturer does not generally have full control over all these aspects, the production-use-reuse chain is more solidly

TABLE 7.1 Manufacturing Sectors and Their Products

Manufacturing Sector	Product Examples	Product Lifetime
Industrial durable goods	Machine tools, motors, fans, air conditioners, conveyer belts, packaging equipment	Very long
Consumer durable goods	Refrigerators, washing machines, furniture, furnaces, water heaters, air conditioners, carpets	Long
Durable medical products	Hospital beds, MRI testing equipment, wheelchairs, washable garments	Long
Vehicles	Automobiles, aircraft, earth movers, snow blowers	Long
Electronics	Computers, cordless telephones, video cameras, television sets, portable sound systems	Moderate
Clothing	Shoes, belts, polyester blouses, cotton pants	Moderate
Consumer nondurable goods	Pencils, batteries, costume jewelry, plastic storage containers, toys	Short
Disposable medical products	Thermometers, blood donor equipment, medicines, non-washable garments	Single use
Disposable consumer products	Antifreeze, paper products, plastic bags, lubricants	Single use
Food products	Frozen dinners, canned fruit, soft drinks, dry cereal	Single use

under his or her purview than under that of the processor, the service provider, or the consumer. Therein lies the manufacturer's great challenge and opportunity.

In devising a streamlined LCA approach, it is important that the full range of environmentally related features of products be encompassed in the analytical tool that is used. The remainder of this chapter presents an example of such a tool.

7.2 MATRIX CONCEPTS

An ideal assessment system for environmentally responsible products (ERPs) should have the following characteristics: It should lend itself to direct comparisons among rated products, be usable and consistent across different assessment teams, encompass all stages of product life cycles and all relevant environmental stressors, and be simple enough to permit relatively quick and inexpensive assessments to be made. Clearly, it must explicitly treat the five life-cycle stages in a typical complex manufactured product, as was illustrated in Figure 2.1.

An assessment system that meets these general criteria was developed at AT&T by the author and his collaborator B.R. Allenby in 1993. It has as its central feature a 5×5 assessment matrix, one dimension of which is life-cycle stages and the other of which is environmental stressors (see Table 7.2). In use, the assessor studies the product design, manufacture, packaging, in-use environment, and likely disposal scenario and assigns to each element of the matrix an integer rating from 0 (highest impact, a very negative evaluation) to 4 (lowest impact, an exemplary evaluation). Because the approach is not quantitative per se, the results are not strictly a measure of environmental performance, but rather an estimate of the potential for improvement in environmental performance. In essence, what the assessor is doing is providing a figure of

TABLE 7.2 The Environmentally Responsible Product Assessment Matrix

Life Stage	Environmental Stressor				
	Materials Choice	Energy Use	Solid Residues	Liquid Residues	Gaseous Residues
Premanufacture	1,1	1,2	1,3	1,4	1,5
Product Manufacture	2,1	2,2	2,3	2,4	2,5
Product Delivery	3,1	3,2	3,3	3,4	3,5
Product Use	4,1	4,2	4,3	4,4	4,5
Refurbishment, Recycling, Disposal	5,1	5,2	5,3	5,4	5,5

(The numbers are the matrix element indices i, j.)

merit to approximate the result of the more formal LCA inventory analysis and impact analysis stages. She or he is guided in this task by experience, a design and manufacturing survey, appropriate checklists, and other information. The process described here is purposely qualitative and utilitarian, but does provide a numerical end point against which to measure improvement.

Although the assignment of integer ratings seems quite subjective, experiments have been performed in which comparative assessments of products are made by several different industrial and environmental engineers. When provided with checklists and protocols (as shown in Appendix A), overall product ratings differ by less than about 15 percent among groups of four assessors.

After an evaluation has been made for each matrix element, the overall Environmentally Responsible Product Rating (R_{ERP}) is computed as the sum of the matrix element values:

$$R_{ERP} = \sum_i \sum_j M_{i,j} \qquad (7.1)$$

Because there are 25 matrix elements, a maximum product rating is 100.

Designers who have never performed a product audit may wonder about the relevance of some of the life stage-environmental stressor pairs. To aid in perspective, Table 7.3 provides examples for each matrix element. The basis for some of these examples is that the industrial process is responsible (implicitly if not explicitly) for the embedded impacts of the processing of raw materials and the use of energy and for the projected impacts as the products are used, recycled, or discarded.

The semiquantitative matrix approach sidesteps several of the problematic aspects of a comprehensive LCA. Defining the functional unit becomes less important, for example: It matters little to an SLCA whether the assessment is being performed from the perspective of a "refrigerator-year" or a "cubic meter of cooled volume-year." Allocation is also avoided, because it is sufficient to identify the major sources and sinks for the principal resources, but not their actual value.

Other important but often confounding LCA components are subsumed in the SLCA checklists and protocols, which become of central importance. It is in these aids that environmental science and industrial ecology are incorporated into the SLCA process, and it is there that such factors as spatial and temporal concerns, localization, and valuation must be taken into account, explicitly or implicitly.

TABLE 7.3 Examples of Product Inventory Concerns

Life Stage	Materials Choice	Energy Use	Solid Residues	Liquid Residues	Gaseous Residues
Premanufacture	Use of only virgin materials	Extraction from ore	Slag production	Mine drainage	SO_2 from smelting
Product Manufacture	Use of only virgin materials	Inefficient motors	Sprue, wrapping disposal	Toxic chemicals	CFC use
Product Delivery	Toxic printing ink use	Energy-intensive packing matls.	Polystyrene packaging	Toxic printing ink use	CFC foams
Product Use	Intentionally dissipated metals	Resistive heating	Solid consumables	Liquid consumables	Combustion emissions
Recycling, Disposal	Use of toxic organics	Energy-intensive recycling	Non-recyclable solids	Non-recyclable liquids	HCl from incineration

Checklists, protocols, and other SLCA guidelines are inherent components of environmental assessment systems throughout industry. They provide the intellectual underpinning that validates a corporation's environmental activities. This crucial role is not widely recognized, however, and the SLCA aids have not yet received the detailed attention they deserve. Accordingly, the appendices in this book should not be regarded as gospel. Rather, they are a first attempt to provide sets of checklists and protocols, and they tend to be general rather than specific. Different industry sectors and different levels of society and government may well develop significant variations applicable to their own goals and scope. The construction and revision of these SLCA aids will be an important activity of the assessment-related efforts of the next decade or more.

7.3 TARGET PLOTS

The matrix displays provide a useful overall assessment of a design, but a more succinct display of DFE design attributes is provided by the *target plots* shown in Figure 7.1. To construct the plots, the value of each element of the matrix is plotted at a specific angle. (For a 25-element matrix, the angle spacing is 360/25 = 14.4°.) A good product or process shows up as a series of dots bunched toward the center, as would occur on a rifle target in which each shot was aimed accurately. The plot makes it easy to single out points far removed from the bulls-eye and to mark their topics out for special attention by the design team. Furthermore, target plots for alternative designs of the same product permit quick comparisons of environmental attributes. The product design team can then select among design options, and can consult the checklists and protocols for information on improving individual matrix element ratings.

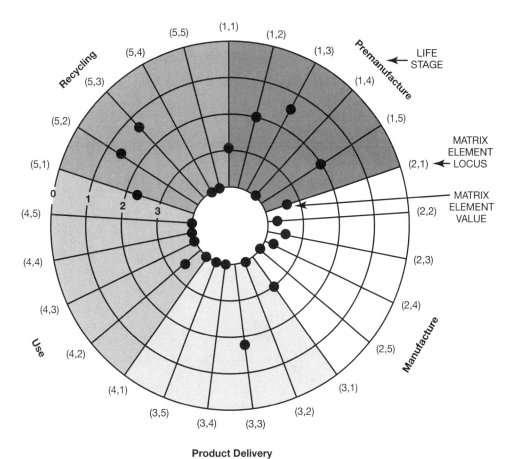

FIGURE 7.1 The features of the target plot.

7.4 ASSESSING GENERIC AUTOMOBILES OF YESTERDAY AND TODAY

As a detailed example of how SLCA is accomplished in practice, the automobile and its manufacture provide a widely known and widely studied product. Automobiles have both manufacturing and in-use impacts on the environment, in contrast to many other products such as furniture or roofing materials. The greatest impacts seem likely to result from the combustion of gasoline and the release of tailpipe emissions during the driving cycle. However, there are other aspects of the product that affect the environment, such as the depletion of fossil fuel resources, the dissipative use of oil and other lubricants, the discarding of tires and other spent parts, and the ultimate retirement of the vehicle. To assess these factors, environmentally responsible product assessments have been performed on generic automobiles of the 1950s and 1990s. As discussed above, the specification of the functional unit is not crucial, but it is something like "one generic automobile, assessed over a spectrum of life stages and environmental stressors." Relevant characteristics of the vehicles are given in Table 7.4. In overview, the 1950s vehicle was substantially heavier, less fuel efficient, prone to

TABLE 7.4 Characteristics of Generic Automobiles (estimates from Ward's Automobile Yearbook)

Characteristic	ca. 1950s Automobile	ca. 1990s Automobile
Materials (kg):		
Plastics	0	101
Aluminum	0	68
Copper	25	22
Lead	23	15
Zinc	25	10
Iron	220	207
Steels	1290	793
Glass	54	38
Rubber	85	61
Fluids	96	81
Other	83	38
Total Weight: (kg)	1901	1434
Fuel Efficiency (miles/gallon)	15	27
Exhaust Catalyst	No	Yes
Air Conditioning	CFC–12*	HFC-134a

*Air conditioning entered the automobile market in the late 1950s on top-of-the-line vehicles.

greater dissipation of working fluids and exhaust gas pollutants, and had components, such as tires, that were less durable.

Premanufacturing, the first life stage, treats impacts on the environment as a consequence of the actions needed to extract materials from their natural reservoirs, transport them to processing facilities, purify or separate them by such operations as ore smelting and petroleum refining, and transport them to the manufacturing facility. Where components are sourced from outside suppliers, this life stage also incorporates assessment of the impacts arising from component manufacture. The ratings assigned to this life stage of generic vehicles from each epoch are given in Table 7.5, and the two numbers in parentheses refer to the matrix element indices. The higher (that is, more favorable) ratings for the 1990 vehicle are mainly due to improvements in the environmental aspects of mining and smelting technologies, improved efficiency of the equipment and machinery used, and the increased use of recycled material.

The second life stage is product manufacture (see Table 7.6). The basic automotive manufacturing process has changed little over the years but much has been done to improve its environmental responsibility. One potentially high-impact area is the paint shop, where various chemicals are used to clean the parts and volatile organic emissions are generated during the painting process. There is now greater emphasis on treatment and recovery of wastewater from the paint shop, and the switch from low-solids to high-solids paint has done much to reduce the amount of material emitted. With respect to material fabrication, there is currently better utilization of material (partially due to better analytical techniques for designing component parts) and a greater emphasis on

TABLE 7.5 Premanufacturing Ratings

Element Designation	Element Value and Explanation
1950s Auto	
Matls. choice (1,1)	2 (Few hazardous materials are used, but most materials are virgin.)
Energy use (1,2)	2 (Virgin material shipping is energy-intensive.)
Solid residue (1,3)	3 (Iron and copper ore mining generate substantial solid residues.)
Liq. residue (1,4)	3 (Resource extraction generates moderate amounts of liquid residues.)
Gas residue (1,5)	2 (Ore smelting generates significant amounts of gaseous residues.)
1990s Auto	
Matls. choice (1,1)	3 (Few hazardous materials are used, and much recycled material.)
Energy use (1,2)	3 (Virgin material shipping is energy-intensive.)
Solid residue (1,3)	3 (Metal mining generates solid residues.)
Liq. residue (1,4)	3 (Resource extraction generates moderate amounts of liquid residues.)
Gas residue (1,5)	3 (Ore processing generates moderate amounts of gaseous residues.)

TABLE 7.6 Product Manufacture Ratings

Element Designation	Element Value
1950s Auto	
Matls. choice (2,1)	0 (Chlorinated solvents, cyanide.)
Energy use (2,2)	1 (Energy use during manufacture is high.)
Solid residue (2,3)	2 (Lots of metal scrap and packaging scrap produced.)
Liq. residue (2,4)	2 (Substantial liquid residues from cleaning and painting.)
Gas residue (2,5)	1 (Volatile hydrocarbons emitted from paint shop.)
1990s Auto	
Matls. choice (2,1)	3 (Good materials choices, except for lead solder waste.)
Energy use (2,2)	2 (Energy use during manufacture is fairly high.)
Solid residue (2,3)	3 (Some metal scrap and packaging scrap produced.)
Liq. residue (2,4)	3 (Some liquid residues from cleaning and painting.)
Gas residue (2,5)	3 (Small amounts of volatile hydrocarbons emitted.)

reusing scraps and trimmings from the various fabrication processes. Finally, the productivity of the entire manufacturing process has been improved, and substantially less energy and time are required to produce each automobile.

The environmental stressors at the third life stage, product packaging and transport, include the manufacture of the packaging material, its transport to the manufacturing facility, residues generated during the packaging process, transportation of the finished and packaged product to the customer, and (where applicable) product installation (see Table 7.7). This aspect of the automobile's life cycle is benign relative to the vast majority of products sold today, because automobiles are delivered with only small amounts of packaging material. Nonetheless, some environmental burden is associated with the transport of a large, heavy product. The slightly higher rating for the 1990s automobile is due mainly to the better design of auto carriers (more vehicles per load) and the increase in fuel efficiency of the transporters.

The fourth life stage, product use, includes impacts from consumables (if any) or maintenance materials (if any) that are expended during customer use (see Table 7.8).

TABLE 7.7 Product Delivery Ratings

Element Designation	Element Value and Explanation
1950s Auto	
Matls. choice (3,1)	3 (Sparse, recyclable materials used during packaging and shipping.)
Energy use (3,2)	2 (Over-the-road truck shipping is energy-intensive.)
Solid residue (3,3)	3 (Small amounts of packaging during shipment could be further minimized.)
Liq. residue (3,4)	4 (Negligible amounts of liquids are generated by packaging and shipping.)
Gas residue (3,5)	2 (Substantial fluxes of greenhouse gases are produced during shipment.)
1990s Auto	
Matls. choice (3,1)	3 (Sparse, recyclable materials used during packaging and shipping.)
Energy use (3,2)	3 (Long-distance land and sea shipping is energy-intensive.)
Solid residue (3,3)	3 (Small amounts of packaging during shipment could be further minimized.)
Liq. residue (3,4)	4 (Negligible amounts of liquids are generated by packaging and shipping.)
Gas residue (3,5)	3 (Moderate fluxes of greenhouse gases are produced during shipment.)

TABLE 7.8 Customer Use Ratings

Element Designation	Element Value and Explanation
1950s Auto	
Matls. choice (4,1)	1 (Petroleum is a resource in limited supply.)
Energy use (4,2)	0 (Fossil fuel energy use is very large.)
Solid residue (4,3)	1 (Significant residues of tires, defective or obsolete parts.)
Liq. residue (4,4)	1 (Fluid systems are very leaky.)
Gas residue (4,5)	0 (No exhaust gas scrubbing; high emissions.)
1990s Auto	
Matls. choice (4,1)	1 (Petroleum is a resource in limited supply.)
Energy use (4,2)	2 (Fossil fuel energy use is large.)
Solid residue (4,3)	2 (Modest residues of tires, defective or obsolete parts.)
Liq. residue (4,4)	3 (Fluid systems are somewhat dissipative.)
Gas residue (4,5)	2 (CO_2, lead [in some locales].)

Significant progress has been made in automobile efficiency and reliability, but automotive use continues to have a very high negative impact on the environment. The increase in fuel efficiency and more effective conditioning of exhaust gases accounts for the 1990s automobile achieving higher ratings, but clearly there is still room for improvement.

The fifth life stage assessment includes impacts during product refurbishment and as a consequence of the eventual discarding of modules or components deemed impossible or too costly to recycle (see Table 7.9). Most modern automobiles are recycled (some 95 percent currently enter the recycling system), and from these, approximately 75 percent by weight is recovered for used parts or returned to the secondary metals market. There is a viable used parts market and most cars are stripped of reusable parts before they are discarded. Improvements in recovery technology have made it easier and more profitable to separate the automobile into its component materials.

In contrast to the 1950s, at least two aspects of modern automobile design and construction are retrogressive from the standpoint of their environmental implica-

Section 7.4 Assessing Generic Automobiles of Yesterday and Today

TABLE 7.9 Refurbishment/Recycling/Disposal Ratings

Element Designation	Element Value and Explanation
1950s Auto	
Matls. choice (5,1)	3 (Most materials used are recyclable.)
Energy use (5,2)	2 (Moderate energy use required to disassemble and recycle materials.)
Solid residue (5,3)	2 (A number of components are difficult to recycle.)
Liq. residue (5,4)	3 (Liquid residues from recycling are minimal.)
Gas residue (5,5)	1 (Recycling commonly involves open burning of residues.)
1990s Auto	
Matls. choice (5,1)	3 (Most materials recyclable, but sodium azide presents difficulty.)
Energy use (5,2)	2 (Moderate energy use required to disassemble and recycle materials.)
Solid residue (5,3)	3 (Some components are difficult to recycle.)
Liq. residue (5,4)	3 (Liquid residues from recycling are minimal.)
Gas residue (5,5)	2 (Recycling involves some open burning of residues.)

tions. One is the increased diversity of materials used, mainly the increased use of plastics but also an increased diversity of alloys. The second aspect is the increased use of welding in the manufacturing process. In the vehicles of the 1950s, a body-on-frame construction was used. This approach was later switched to a unibody construction technique in which the body panels are integrated with the chassis. Unibody construction requires about four times as much welding as does body-on-frame construction, plus substantially increased use of adhesives. The result is a vehicle that is stronger, safer and uses less structural material, but is much less easy to disassemble.

The completed matrices for the generic 1950s and 1990s automobile are illustrated in Table 7.10. Examine first the values for the 1950s vehicle so far as life stages are concerned. The column at the far right of the table shows moderate environmental stewardship during resource extraction, packaging and shipping, andrefurbishment/recycling/disposal. The ratings during manufacturing are poor, and during customer

TABLE 7.10 Environmentally Responsible Product Assessments for the Generic 1950s and 1990s Automobiles*

	Environmental Stressor					
Life Stage	Materials Choice	Energy Use	Solid Residues	Liquid Residues	Gaseous Residues	Total
Premanufacture	2	2	3	3	2	12/20
	3	3	3	3	3	15/20
Product Manufacture	0	1	2	2	1	6/20
	3	2	3	3	3	14/20
Product Delivery	3	2	3	4	2	14/20
	3	3	3	4	3	16/20
Product Use	1	0	1	1	0	3/20
	1	2	2	3	2	10/20
Refurbishment, Recycling, Disposal	3	2	2	3	1	11/20
	3	2	3	3	2	13/20
Total	9/20	7/20	11/20	13/20	6/20	46/100
	13/20	12/20	14/20	16/20	13/20	68/100

*Upper numbers refer to the 1950s automobile, lower numbers to the 1990s automobile.

Chapter 7 Product Assessment by SLCA Matrix Approaches

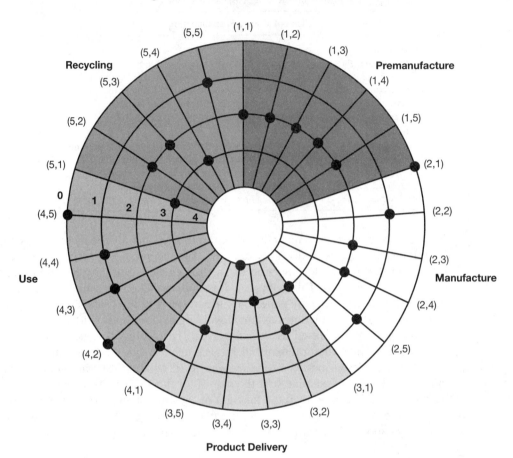

FIGURE 7.2 Comparative target plots for the display of the environmental impacts of the generic automobile of the 1950s and of the 1990s.

use are abysmal. The overall rating of 46 is far below what might be desired. In contrast, the overall rating for the 1990s vehicle is 68, much better than that of the earlier vehicle but still leaving plenty of room for improvement. A more succinct display of DFE design attributes is provided by the target plots of Figure 7.2.

7.5 DISCUSSION

The product assessment system presented above can be readily adapted to the manufacture of a variety of products. In cases where a corporation's product is another corporation's feedstock, as is the case with producers of plastic pellets eventually used for auto body panels, for example, assessments can be (and have been) done on an intercorporate basis.

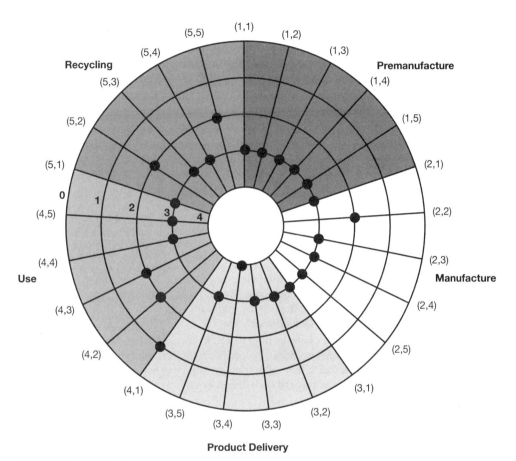

FIGURE 7.2 Continued.

Unlike classical inventory analysis and perhaps impact analysis, overall life-cycle assessment as presented here is less quantifiable and less thorough. It is also inestimably more practical and utilitarian; it is far better to conduct a number of streamlined LCAs by these or similar techniques than to conduct one or two comprehensive LCAs. This is particularly true because LCAs and SLCAs are about equally successful at identifying actions that will improve environmental performance, because the options of designers are generally constrained to a fairly high degree, and because the results of full-scale LCAs remain contentious. An SLCA survey of the modest depth advocated here, performed by an objective professional or a cross-functional group of professionals, will succeed in identifying perhaps 80 percent of useful design actions that could be taken, while consuming sufficiently small levels of time and money. Furthermore, that assessment has a good chance of being carried out and its recommendations of being implemented.

FURTHER READING

Field F., J. Ehrenfeld, D. Roos, and J. Clark. *Automobile Recycling Policy: Findings and Recommendations*, Prepared for the Automotive Industry Board of Governors (Cambridge, MA: Massachusetts Institute of Technology, February, 1994).

Graedel, T.E., B.R. Allenby, P. Comrie, Matrix approaches to abridged life cycle assessment, *Environmental Science & Technology, 29* (1995) 134A–139A.

Henstock, M.E., *Design for Recyclability* (London: The Institute of Metals, 1988).

Noyes, R., Ed., *Pollution Prevention Technology Handbook* (Park Ridge, NJ: Noyes, 1993).

U.S. Congress, Office of Technology Assessment, *Green Products by Design: Choices for a Cleaner Environment*, Rpt. OTA-E-541 (Washington, DC: U.S. Govt. Printing Office, 1992).

EXERCISES

7.1. Choose five manufactured items of different types that you come into contact with on a regular basis. For each, list the major materials from which it is made, the original source of those materials, and the typical lifetime of the product. Is it likely or possible that the materials will be recycled?

7.2. Five environmental stressors form one axis of the assessment matrix of Table 7.2. Consider the environmental concerns of Table 1.3 and the Targeted Activities of Table 1.4. Should different stressors or additional stressors be included in the generalized assessment matrix? Defend your answer.

7.3. Product inventories have been prepared for two different designs of a high speed widget. The matrices are reproduced below, the figure on the left side of each matrix element referring to Design 1, that to the right to Design 2. Select the better product from a DFE viewpoint. What features of each design would you address if improvement were needed?

Life Stage	MC	EU	SR	LR	GR
PM	1/1	4/3	4/3	2/2	3/2
M	2/1	1/2	1/2	2/1	2/4
PD	3/2	1/1	2/3	1/1	1/1
PU	1/2	1/2	1/3	1/1	1/3
RD	2/1	2/2	2/1	1/2	1/2

7.4. Draw target plots for the two matrices of Exercise 7.3.

CHAPTER 8
Process Assessment by SLCA Matrix Approaches

8.1 CONSIDERATIONS IN PROCESS ASSESSMENT

Industrial processes have substantial inertia, much more so than manufactured products. The capital and personnel costs of installing processes are high, and they often remain in place for decades. The costs of modifying or retrofitting them is high as well, and processes installed without much attempt to take long-term environmental considerations into account may eventually require substantial investments in pollution control add-ons and treatment facilities. It is thus very important, from both operational and environmental standpoints, to "do it right the first time."

To begin to think about processes, consider one of the simplest and most common: the preparation of food by cooking. As shown in Figure 8.1a, the purpose of the process is to transform raw food (the incoming material) into cooked food (the outgoing product). Doing so requires energy and perhaps some process chemicals such as cooking oil. Residues of the process not retained in the product are waste heat, food waste (food not fully removed from the cooking equipment, discarded cooked food, and so forth), and gaseous emissions.

Most food that is cooked is not provided to the cooking process exactly as it is received from the grocer. More commonly, washing and trimming is required. This is a process complementary to the primary cooking process; that is, the choice of the cooking process implies the simultaneous provisioning of a washing and trimming process. In that latter process, shown in Figure 8.1b, the process chemical is water, the residues are wastewater and food waste, the input material is unprepared raw food, and the product is prepared raw food.

The choice of the cooking process not only implies that a complementary process is available, but also that the equipment used in each process be chosen, purchased, installed, and, eventually, discarded. Thus, one would consider the environmental implications of the gas or electric range itself, the sink used for washing and trimming, and the necessary piping and plumbing. Two aspects of the SLCA not encountered in product SLCAs are involved here: the consideration of capital goods and the extension of the analysis to time scales of decades. These aspects add important perspective to the assessment, but add complexity as well.

Assessing a process can take place at several levels. The simplest and generally most tractable is to construct flow diagrams like those of Figure 8.1 and evaluate the

112   Chapter 8   Process Assessment by SLCA Matrix Approaches

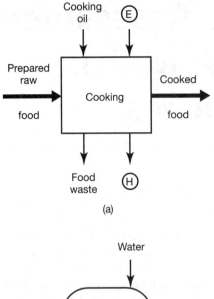

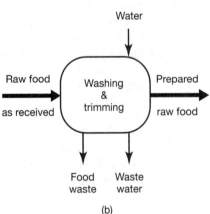

FIGURE 8.1 Representations of flows of materials and energy for the processes used in the cooking of food (a) and the washing and trimming of food (b). Raw materials enter the process from the left and products leave from the right. A circled E indicates energy input, a circled H indicates heat output. Process materials enter from the top and waste products leave from the bottom. Heavy lines indicate flows of products, lighter lines flows of process resources. The circled E indicates the use of energy and the circled H indicates the loss of energy as heat.

inputs and outputs. The next stage that can be considered is to move one level back from that and consider that the use of cooking oil implies the processing of olive oil, the use of energy implies emissions at the electrical power plant, and the use of water involves consideration of the environmental impacts of the water storage, treatment, and delivery system.

8.2 STAGES IN PROCESS LIFE CYCLES

As with products, industrial processes can be evaluated by SLCA matrix techniques simple enough to permit relatively quick and inexpensive assessments to be made, and in which all stages of product life cycles and all relevant environmental stressors are encompassed. The assessment system for products presented in Chapter 7 can be readily adapted to these requirements. As before, it features a 5×5 matrix, one dimension of which is life-cycle stages and the other of which is environmental stressors (Table 8.1). The environmental stressors are the same as those for products, but the life-cycle

TABLE 8.1 The Environmentally Responsible Process Matrix

Life Stage	Environmental Stressor				
	Materials Selection	Energy Use	Solid Residues	Liquid Residues	Gaseous Residues
Resource Provisioning	1,1	1,2	1,3	1,4	1,5
Process Implementation	2,1	2,2	2,3	2,4	2,5
Primary Process Operation	3,1	3,2	3,3	3,4	3,5
Complementary Process Operation	4,1	4,2	4,3	4,4	4,5
Refurbishment, Recycling, Disposal	5,1	5,2	5,3	5,4	5,5

(The numbers are the matrix element indices i, j.)

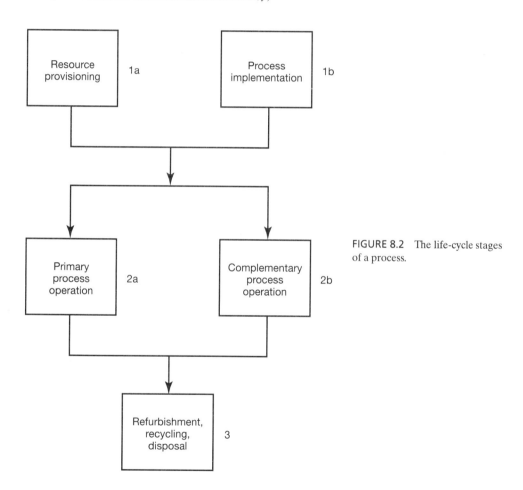

FIGURE 8.2 The life-cycle stages of a process.

stages are different. Unlike product life stages, which are sequential, process life stages have only three epochs (Figure 8.2): resource provisioning and process implementation occur simultaneously, primary process operation and complementary process operation occur simultaneously as well, and refurbishment, recycling, and disposal is the end-of-life stage. The characteristics of these life stages are described below.

8.2.1 Resource Provisioning

The first stage in the life cycle of any process is the provisioning of the materials used to produce the consumable resources that are used throughout the life of the product being assessed. One consideration is the source of the materials, which in many cases will be extracted from their natural reservoirs. Alternatively, recycled materials may be used. Doing so may be preferable to using virgin materials because recycled materials avoid the environmental disruption that virgin material extraction involves, and often require less energy to recover and recycle than would be expended in virgin material extraction. In addition, the recycling of materials often produces less solid, liquid, or gaseous residues than do virgin materials extractions. Tradeoffs, especially in energy use, are always present, however, and the choice must be considered on a case-by-case basis. Another consideration is the methods used to prepare the materials for use in the process. Regardless of the source of a metal sheet to be formed into a component, for example, the forming and cleaning of the sheet and the packaging of the component should be done in an environmentally responsible manner. Supplier operations are thus a topic for evaluation as the process is being developed and, later, as it is being used.

8.2.2 Process Implementation

Coincident with resource provisioning is process implementation, which looks at the environmental impacts that result from the activities necessary to make the process happen. These principally involve the manufacture and installation of the process equipment, and installing other resources that are required such as piping, conveyer belts, exhaust ducts, and the like. This row of the matrix has a strong commonality with the Chapter 7 matrix for evaluating a product.

8.2.3 Primary Process Operation

A process should be designed to be environmentally responsible in operation. Such a process would ideally limit the use of hazardous materials, minimize the consumption of energy, avoid or minimize the generation of solid, liquid, or gaseous residues, and ensure that any residues that are produced can be used elsewhere in the economy. Effort should be directed to designing processes whose secondary products are salable to others or usable in other processes within the same facility. In particular, the generation of residues whose hazardous properties render their recycling or disposal difficult should be avoided. Because successful processes can become widespread throughout a manufacturing sector, they should be designed to perform well under a variety of conditions.

An unrealizable goal—but a useful target—is that every molecule that enters a manufacturing process should leave that process as part of a salable product. One's intuitive perception of this goal as unrealistic is not necessarily accurate: Certain of today's manufacturing processes, such as molecular beam epitaxy, come close, and more will do so in the future.

8.2.4 Complementary Process Operation

It is often the case that several manufacturing processes form a symbiotic relationship, each assuming and depending on the existence of others. Thus, a comprehensive process evaluation needs to consider not only the environmental attributes of the primary process itself, but also those of the complementary processes that precede and follow. For example, a welding process generally requires a preceding metal cleaning step, which traditionally required the use of ozone-depleting chlorofluorocarbons. Similarly, a soldering process generally requires a post-cleaning to remove the corrosive solder flux. This step also traditionally required the use of chlorofluorocarbons. Changes in any element of this system—flux, solder, or solvent—usually require changes to the others as well if the entire system is to continue to perform satisfactorily. The responsible primary process designer will consider to what extent his process imposes environmentally difficult requirements for complementary processes, both in their implementation and their operation.

8.2.5 Refurbishment, Recycling, Disposal

The process designer must recognize that all process equipment will eventually become obsolete, and that it must therefore be designed to optimize disassembly and reuse, either of modules (the preferable option) or materials. In this sense, process equipment is subject to the same considerations and recommended activities that apply to any product: use of quick disconnect hardware, identification marking of plastics, and so on. Many of these design decisions are made by the corporation actually manufacturing the process equipment, which may well not be the user, but the process designer can control or frustrate many environmentally responsible equipment recycling actions by his or her choice of features on the original process design.

A classic example of the consequences of failing to design for recycling is the Brent Spar oil platform. This North Sea installation was designed and built in the 1960s as a temporary storage reservoir for crude oil awaiting transfer to refineries. In 1995, Shell Oil, after a careful review of alternatives, decided to dispose of the then-obsolete platform by scuttling it in the Atlantic Ocean well off the west coast of Scotland. This action was seen by Greenpeace, other environmental organizations, and most of Western Europe's citizens as personifying the wasteful society and the disregard of industry for the environment. For several weeks, as Shell Oil debated options and Greenpeace boats circled the platform, Shell gasoline sales dropped sharply. Finally giving in to the pressure, Shell agreed to bring the platform to shore and pay large sums to have it cut up, the petroleum residues properly treated, and the metal recycled. It made little difference that the overall environmentally preferable choice (if energy, exposure to hazardous material, and use of recycled material were considered) may well have been the scuttling originally proposed. The fact was that Shell had paid an enormous price, both monetarily and from a public-relations point of view, for never considering the final life stage in the design of the platform.

8.2.6 The Rating Matrix

In arriving at an individual matrix element assessment for processes, or in offering advice to designers seeking to improve the rating of a particular matrix element, the assessor can refer for guidance to underlying checklists and protocols. (An example of the contents of elements of such checklists and protocols for processes is given in Appendix B.) After an evaluation has been made for each matrix element, the overall environmentally responsible process rating is computed, as in the case of products, as the sum of the matrix element values:

$$R_{ERPS} = \sum_i \sum_j M_{i,j} \qquad (8.1)$$

Because the process matrix has 25 matrix elements, each with a maximum value of 4, the maximum process assessment rating is 100.

8.3 THE APPROACH TO PROCESS ANALYSIS

Process analysis begins with data gathering, and the focus is in three areas: the primary process itself, the equipment used in the process, and information concerning any complementary processes involved. The information should be as quantitative as possible, provided quantification can be done quickly, but much can be done with qualitative data and the analyst should not get bogged down in attempting to precisely quantify everything.

After the data are assembled, the checklist in Appendix B can be used in making matrix element assessments. As before, the goal is to affect improvements, not to get mired in the quicksand of attempting to do a perfect job of matrix element evaluation. A list of recommendations should then be generated and those recommendations prioritized by Pareto plot techniques, just as was done for the environmentally responsible product assessment shown in Chapter 7.

8.3.1 The Process Itself

The analysis of a process begins with the study of the actual operation that the process performs. Almost certainly, the process requires energy. What is the energy source, and can less energy or a more benign source of energy be used? Are energy recovery or cogeneration used? Often the process requires chemicals. If so, what are the chemicals used, and are they hazardous to humans, other biological species, or ecosystems? Are the chemicals from virgin sources or from recycling streams? If they are from the former, what recycling streams or outputs might be available for use?

Output streams also require analysis. What are the chemical by-products of the process, if any? (Note that by-product production can occur without an input chemical stream because the by-product may be derived from the incoming component, as in turnings from lathe operation.) If energy is consumed, it is very likely that heat is given off. Is the process well insulated so that little of that heat is lost? Alternatively, is the utility of the heat captured in any way and reused, say to heat nearby offices? (It is worth pointing out that most households, offices, and light commercial operations avoid industrial areas. In fact, zoning laws may dictate separation. Thus, the best potential customers for low-pressure steam or heated water may be excluded from consideration by geography and public policy.)

The mechanical and relational arrangements of the processes can also be a useful item to review. Is the process batch or continuous? If it is batch, are energy and materials requirements for startup minimized? Is the process located in proximity to other processes or to flows of incoming or outgoing components so as to minimize transport requirements? If a substantial by-product stream is generated, is there a nearby process that can receive and use it (either within the corporation or in another nearby corporation)?

The information derived from answering these questions permits the analyst to complete rows one and three of the process matrix.

8.3.2 The Process Equipment

Process equipment should be analyzed as though one were analyzing a product. (It is, of course, the product of the process equipment manufacturer, and the purchaser of the process equipment assumes the equipment's environmental attributes, good and bad.) Using the guidelines in Chapter 7, study the materials used to manufacture the equipment, the methods by which the equipment is assembled, its modularity, the ease with which the equipment can be disassembled, and the degree to which the materials are identified. Most process equipment is made from steel. Depending on the process, stainless steel or organic or ceramic surface coatings may be required. Process equipment is also generally painted on exterior surfaces. Some process equipment, especially outside the heavy machinery industries, includes or is made entirely of plastic. Computer control, and the associated electronics components, are common.

If the process is being installed or soon will be, examine the techniques and materials used for packaging, shipping, and installation.

Energy use is an important attribute of a product made to be used in a manufacturing process. Any component that draws current should be designed to be capable of shutdown when not in active use. Motors should generally be the variable-speed type, the speed being load controlled.

The information derived from answering these questions permits the analyst to complete rows two and five of the process matrix. If the process is not already in place, but is being designed or will soon be installed, the analyst will have a particularly complete picture of the row two information.

8.3.3 Complementary Processes

Examine whether the primary process being assessed requires preceding or subsequent processes, and whether those processes are of a particular type or use a particular chemical. If the complementary processes are thus defined, is the primary process itself environmentally responsible? If not, can it be modified to improve its characteristics?

If any of several complementary processes can be used, have the designers chosen ones that are environmentally responsible? If not, can alternatives be suggested?

The information derived from answering these questions permits the analyst to complete row four of the process matrix.

8.4 ASSESSING GENERIC AUTOMOBILE MANUFACTURING PROCESSES

As a demonstration of the process assessment tool described above, it is instructive to perform environmentally responsible process assessments on common processes that can be readily visualized, at least in concept: those utilized in generic automobile manufacturing plants of the 1950s and 1990s. The functional unit here is one generic automobile manufacturing facility, assessed over a spectrum of life stages and environmental stressors. Some of the relevant characteristics of the processes are given in Table 8.2, and process sequence diagrams for the two epochs are shown in Figure 8.3. In the 1950s, the main automobile components were made of steel, and the processes were primarily those that cast and worked steel: sand casting, metal forming, welding, plating, and painting. In the 1990s, aluminum was incorporated as a primary material for automobile frames and for some sheet metal and components. This change added die casting to the list of processes, largely replacing the metal plating process that had widely been minimized. In both epochs, the primary manufacturing processes chosen implied the coexistence of at least two complementary processes: trimming and smoothing of castings, and metal cleaning.

The processes used for manufacture in the 1950s were much more energy-intensive than is the modern practice, and some problematic processes, such as metal cleaning with chlorofluorocarbons, plating with cyanides, and painting with high-volatility organics, have since been modified or eliminated. The internal recycling of materials, once hardly considered, now occurs much more extensively.

8.4.1 Primary Process Descriptions

While it is fairly straightforward to picture and describe the processes by which typical automobiles of different eras were made, there has historically been some significant changes in the types of processes used in that manufacture, as well as in the ways in which those processes were implemented. Thus, for pedagogical purposes, the two extremes described here will serve to illustrate assessment ranking differences.

8.4.1.1 Sand Casting

Casting is the process of forming objects by pouring or injecting liquid into a mold. Components of steel that require detailed shapes and patterns are generally manufactured by sand casting. In this process, selected and prepared sand is packed into a steel mold housing, and the reverse image of the component is formed in sand coated with an organic binder (generally by making an image of the component to be manufactured during casting), forming the mold around the image, and then removing the image. Molten steel is then poured into the mold and allowed to harden. The sand is brushed away from the component, which can then be cleaned and further processed.

TABLE 8.2 Salient Characteristics of Processes for Generic Automobile Manufacturing of the 1950s and 1990s.

Characteristic	ca. 1950s	ca. 1990s
Sand Casting		
Implementation	Ubiquitous	Ubiquitous
Matl. source	Virgin	Some recycled
Energy use	Very high	High
Residues	Contaminated sand	Contaminated sand
Die Casting		
Implementation	No	Ubiquitous
Matl. source	N/A	Some recycled
Energy use	N/A	High
Residues	N/A	Wastewater
Metal Forming		
Implementation	Ubiquitous	Ubiquitous
Matl. source	Virgin	Some recycled
Energy use	Very high	High
Residues	Trimmings, lubricants	Trimmings, lubricants
Welding		
Implementation	Ubiquitous	Ubiquitous
Matl. source	Virgin	Some recycled
Energy use	High	Moderate
Residues	Negligible	Negligible
Metal Plating		
Implementation	Ubiquitous	Common
Matl. source	Virgin	Some recycled
Energy use	Moderate	Moderate
Residues	Toxic wastewater	Toxic wastewater
Painting		
Implementation	Ubiquitous	Ubiquitous
Matl. source	Virgin	Some recycled
Energy use	Moderate	High
Residues	Much VOC	Minor VOC
Trimming and Smoothing of Metal		
Implementation	Ubiquitous	Ubiquitous
Matl. source	Virgin	Some recycled
Energy use	Very high	High
Residues	Metal trimmings	Metal trimmings
Metal Cleaning		
Implementation	Ubiquitous	Ubiquitous
Matl. source	Virgin	Some recycled
Energy use	Moderate	Moderate
Residues	CFCs	Aqueous

Transforming the metal from the solid to the molten state requires substantial energy, much of which is eventually lost as heat. The organic binder on the loose sand undergoes chemical reactions in the casting process, and after casting it contains potentially hazardous materials such as polynuclear aromatic hydrocarbons. As a consequence, and because cleaning the sand is expensive and new sand is cheap, the used

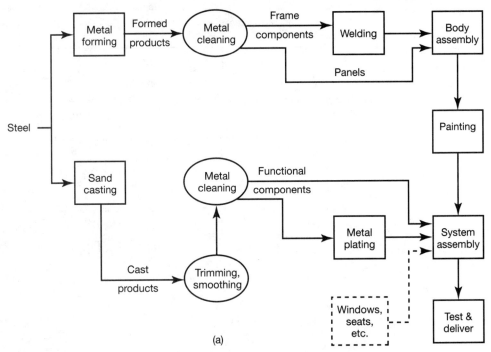

FIGURE 8.3 Process sequence diagrams for generic automobile manufacture in the 1950s (a) and 1990s (b). Primary processes appear in rectangles, complementary processes in ovals. The boxes in dashed lines are included to indicate that only the principal metal structural components and their processes are treated, but that a more comprehensive analysis would include other components and their processes as well.

foundry sands are discarded as waste. The flow diagram of materials, energy, and components for sand casting is shown in Figure 8.4a.

8.4.1.2 Die Casting

The manufacture of aluminum automotive parts is generally carried out by die casting. In this operation, a die of tool steel forms the reverse image of the desired component. Molten aluminum is forced under pressure into the die and allowed to harden. An organic "parting agent" is sprayed onto the die surface prior to pouring to keep the molten aluminum from welding itself to the die. After each part is molded, the die must be cooled by cold water spray from a temperature of more than 1,000°C to less than 400°C. The spray also cleans degraded parting agent from the die surface, which is then recoated before the next casting is made.

As with steel, transforming aluminum from the solid to the molten state requires substantial energy, much of which is eventually lost as heat. In addition, the water used to cool the die becomes contaminated by the parting agent and must undergo a purifi-

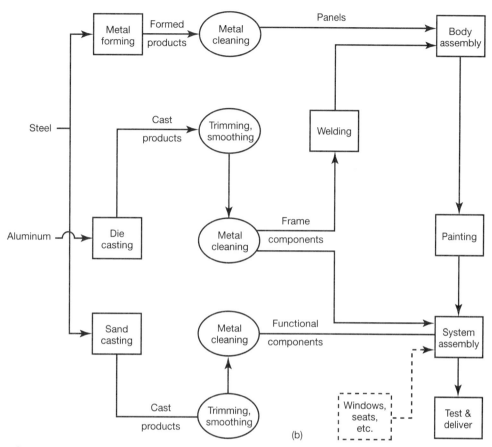

FIGURE 8.3 Continued.

cation step before reuse; alternatively, it is discarded. The flow diagram of materials, energy, and components for die casting is shown in Figure 8.4b.

8.4.1.3 Metal Forming

Forming is a process by which the size or shape of a metal part is changed by producing stresses in the part that exceed the yield strength but not the fracture strength. There are many varieties of forming, including pressing, hammering, rolling, and drawing; most are used in the manufacture of some components of the typical automobile, such as body panels and engine, transmission, and suspension parts. The metal forming equipment that contacts the parts is of tool steel, while the remainder of the equipment is generally of carbon steel.

Forming processes are energy-intensive, and significant heat is produced as a by-product. Lubricants are often utilized to enhance machinability, thus requiring the purification or disposal of contaminated lubricants. Additionally, most forming processes are followed by a trimming stage to remove excess metal; the trimming

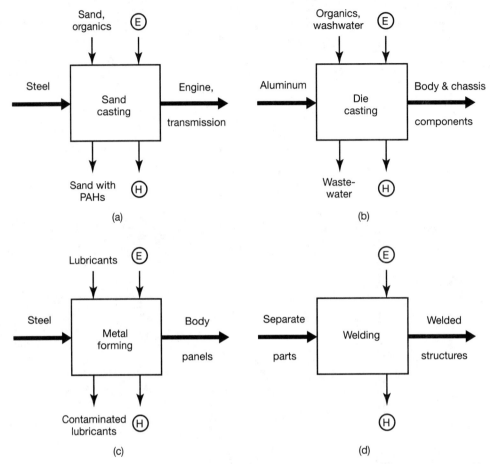

FIGURE 8.4 Representations of flows of materials, energy, and components for primary processes used in automobile manufacture. (a) Sand casting; (b) Die casting; (c) Metal forming; (d) Welding; (e) Metal plating; (f) Painting. The symbols are as in Figure 8.1. The list of processes is intended to be illustrative, not comprehensive.

scraps must then be recycled in some fashion. The flow diagram of materials, energy, and components for metal forming is shown in Figure 8.4c.

8.4.1.4 Welding

The process of welding is one in which metal parts are joined by applying heat, and sometimes pressure, at the joint between the parts. Generally, a metal is added to fill in the joint being melted. In electric-arc and gas-torch welding, the filler metal is the same as the base metal. In brazing and soldering, the filler metal (which has a lower melting point) is melted, but the base metal is not.

Welding requires substantial amounts of energy to melt the filler metal (and often the base metal). No residues of consequence are produced. The flow diagram of materials, energy, and components for welding is shown in Figure 8.4d.

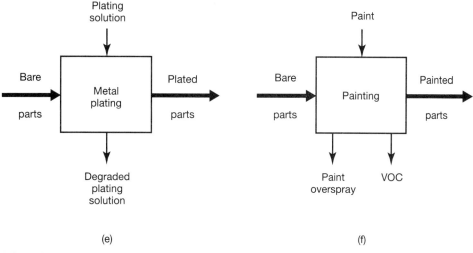

FIGURE 8.4 Continued.

8.4.1.5 Metal Plating

Plating is the process of depositing a thin layer of one metal on another, generally for protective and/or decorative purposes. In electroplating, the piece to be plated is immersed in a solution and made the cathode of a direct current circuit. The coating metal acts as the anode, replenishing the solution as its ions are attracted to the piece being plated. In electroless plating, the metal is deposited from solution without an imposed current being applied. Plating tanks are generally of carbon steel, and may be lined with an organic corrosion-protection layer.

Most plating solutions and most plating metals that are highly protective are also highly toxic. As a consequence, metal plating has rapidly decreased in frequency of use in recent years, and parts that formerly were of plated metal are often replaced with parts formed from plastics or composites. The flow diagram of materials, energy, and components for metal plating is shown in Figure 8.4e.

8.4.1.6 Painting

The painting operation (or, more generally, the surface coating operation) is performed on essentially all visible exterior and interior body components on all products. Painting involves the application of pigment in a solvent carrier. The carrier has traditionally been organic in nature, although some current carriers are aqueous solvents.

Environmentally sound painting operations require the capture and disposal of paint overspray and the containment of greater or lesser quantities of volatilized paint and carrier. Because of the large quantities of air used in modern painting systems, the energy consumed for blowers and filters is large. The flow diagram of materials, energy, and components for painting is shown in Figure 8.4f.

8.4.2 Complementary Process Descriptions

In the automobile manufacturing scenario as outlined here, there are two complementary processes: trimming and smoothing of castings, and metal cleaning. They are described below.

8.4.2.1 Trimming and Smoothing of Castings

After castings are removed from the molds in which they are made, there is invariably excess metal (called "sprues" and "runners") that results from mold seams, pouring spouts, and the like. This excess material must be trimmed away and the metal then smoothed so that it will fit properly and have a satisfactory appearance.

Trimming and smoothing are not energy-intensive, but normally involve the use of tool-steel machinery to cut and abrade. Organic lubricants are commonly used. The flow diagram of materials, energy, and components for trimming and smoothing is shown in Figure 8.5a.

8.4.2.2 Metal Cleaning

During casting, forming, and other process operations, metal parts tend to collect dirt, oil, and other contaminants, which must be removed prior to surface coating or other contaminant-sensitive operations. The cleaning step generally involves the use of liquid solvents, organic and inorganic, in a lined steel tank, often followed by an aqueous rinsing stage.

For many years, chlorinated compounds were the solvents of choice for metal cleaning as a consequence of their nonflammability and satisfactory cleaning capability. With the discovery of their carcinogenic effects, as well as the impacts of some of

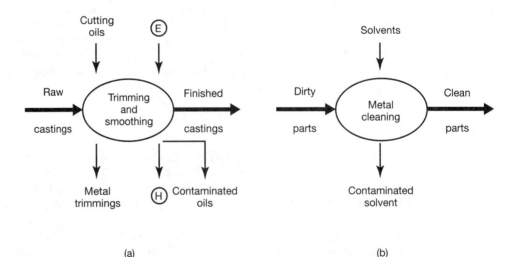

FIGURE 8.5 Representations of flows of materials, energy, and components for complementary processes used in automobile manufacture. (a) Trimming and smoothing of castings; (b) Metal cleaning. The symbols are as in Figure 8.1.

them on the stratospheric ozone layer, the chlorinated solvents have generally been replaced by hydrogenated variants (HCFCs) or alkaline solutions. No matter what the cleaning solvent, contaminated solvent is the inevitable result of metal cleaning processes; the solvent must then be purified in order to be reused. Alternatively, it can be disposed of in a suitable manner. The flow diagram of materials, energy, and components for metal cleaning is shown in Figure 8.5b.

8.4.3 The Process Assessment

The assessment is begun with life stage 1a, resource provisioning, and is guided by the matrix element checklists. An obvious difference between the two epochs is that almost no recycled materials were used in the 1950s, and substantial amounts were used in the 1990s. This has impacts on three environmental stresses: energy use (less is needed to recycle material than to mine and process virgin material), solid residues (no rock overburden results from the use of recycled materials), and gaseous residues (modest emissions result from materials reprocessing as opposed to the initial smelting process). Neither particularly hazardous or otherwise undesirable materials are used in either era, nor are liquid residues a significant problem. The ratings, and brief descriptions of the principal reasons for them, are shown in Table 8.3.

Life stage 1b is primary process implementation, in which the manufacture of the process equipment is evaluated (Table 8.4). Equipment manufacturers in the 1950s used chlorinated solvents extensively for metal cleaning; by the 1990s this use was uncommon or had ceased. Energy use in equipment manufacturing had also declined significantly. Equipment manufacture generated significant solid, liquid, and gaseous emissions in the 1950s and much reduced emissions in the 1990s.

Life stage 2a is primary process operation (Table 8.5). The 1950s ratings reflect high emissions of toxic species and profligate energy use. Improvements had been made by the 1990s, especially the elimination of most metal plating except for zinc anticorrosion treatment of body panels, but the disposal of casting sand and significant use of energy and organic solvents remain as challenges.

TABLE 8.3 Resource Provisioning

Element Designation	Element Value and Explanation
1950s Processes	
Matls. selec. (1,1)	2 (Virgin material use)
Energy use (1,2)	0 (Virgin materials acquisition)
Solid residue (1,3)	1 (Virgin materials acquisition)
Liq. residue (1,4)	3 (Heavy metal leachate)
Gas residue (1,5)	1 (Virgin materials acquisition)
1990s Processes	
Matls. selec. (1,1)	3 (Some virgin materials use)
Energy use (1,2)	2 (Partially recycled materials)
Solid residue (1,3)	2 (Moderate mining residues)
Liq. residue (1,4)	4 (No significant concerns)
Gas residue (1,5)	3 (Smelting emissions are modest)

TABLE 8.4 Process Implementation

Element Designation	Element Value and Explanation
1950s Processes	
Matls. selec. (2,1)	0 (CFCs used for cleaning)
Energy use (2,2)	0 (Very high energy use)
Solid residue (2,3)	2 (Moderate packaging residues)
Liq. residue (2,4)	2 (Plating and trimming effluents)
Gas residue (2,5)	2 (Substantial gaseous emissions)
1990s Processes	
Matls. selec. (2,1)	3 (Nontoxic cleaning solvents)
Energy use (2,2)	2 (Variable speed motors, other conservation)
Solid residue (2,3)	2 (Moderate packaging residues)
Liq. residue (2,4)	4 (No significant concerns)
Gas residue (2,5)	3 (Modest gaseous emissions)

TABLE 8.5 Primary Process Operation

Element Designation	Element Value and Explanation
1950s Processes	
Matls. selec. (3,1)	0 (Hazardous plating solutions)
Energy use (3,2)	1 (Very high energy use)
Solid residue (3,3)	1 (Casting sand residues)
Liq. residue (3,4)	1 (Casting wastewater residues)
Gas residue (3,5)	1 (VOCs from painting)
1990s Processes	
Matls. selec. (3,1)	2 (Organic solvents)
Energy use (3,2)	2 (Significant energy use)
Solid residue (3,3)	1 (Casting sand residues)
Liq. residue (3,4)	3 (Most wastewater recycled)
Gas residue (3,5)	3 (Most VOC emissions captured)

Life stage 2b is complementary process operation. Metal trimming and smoothing has not changed greatly over 40 years, but process energy use in the 1990s is less than in the 1950s, and the trimmings and filings are better contained. A major improvement in metal cleaning is the transition from chlorinated solvents to organics, in many cases to aqueous cleaners, and in some cases to no cleaning at all. The ratings for the two generic sets of processes are shown in Table 8.6.

The final life stage is process refurbishment, recycling, or disposal. Solid residue recovery in 1990s process equipment is improved over the 1950s, although additional energy is needed to accomplish the recovery. The ratings are shown in Table 8.7.

The completed matrices for the generic 1950s and 1990s automobile manufacturing processes appear in Table 8.8. Examine first the values for the 1950s processes so far as life stages are concerned. The column at the far right of the table shows moderate environmental stewardship during the final life stage. The ratings during resource provisioning are poor, and during process implementation and primary and comple-

Section 8.5 Conducting the Assessment and Prioritizing the Recommendations 127

TABLE 8.6 Complementary Process Operation

Element Designation	Element Value and Explanation
1950s Processes	
Matls. selec. (4,1)	0 (Chlorinated solvents for cleaning)
Energy use (4,2)	2 (Moderate energy for trimming and smoothing)
Solid residue (4,3)	2 (Metal trimmings)
Liq. residue (4,4)	0 (Disposal of contaminated solvent)
Gas residue (4,5)	1 (Uncontrolled volatile emissions)
1990s Processes	
Matls. selec. (4,1)	3 (Nontoxic solvents)
Energy use (4,2)	3 (Modest energy for trimming and smoothing)
Solid residue (4,3)	3 (Modest metal trimmings)
Liq. residue (4,4)	2 (Recycling of contaminated solvent)
Gas residue (4,5)	3 (Minimal volatile emissions)

TABLE 8.7 Refurbishment/Recycling/Disposal

Element Designation	Element Value and Explanation
1950s Processes	
Matls. selec. (5,1)	3 (Some hazardous platings)
Energy use (5,2)	3 (Some energy used to recover steel)
Solid residue (5,3)	1 (Partial recovery of steel)
Liq. residue (5,4)	2 (Plating and solvent residues)
Gas residue (5,5)	4 (No significant concerns)
1990s Processes	
Matls. selec. (5,1)	4 (Most materials recyclable)
Energy use (5,2)	1 (High energy use in metals recovery)
Solid residue (5,3)	3 (Only modest design for disassembly)
Liq. residue (5,4)	4 (No significant concerns)
Gas residue (5,5)	4 (No significant concerns)

mentary process operation the ratings are abysmal. The overall rating of 38 is far below what might be desired. In contrast, the overall rating for the 1990s vehicle manufacturing processes is 70, with all life stages being improved. This is much better than that of the earlier era but still leaves plenty of room for improvement.

The target plots for the manufacturing processes are shown in Figure 8.6. Here the lack of change in the rating of life stage 3 and the great changes over time in life stages 1b, 2a, and 2b are obvious.

8.5 CONDUCTING THE ASSESSMENT AND PRIORITIZING THE RECOMMENDATIONS

Environmentally responsible planning for processes, especially new ones, is a four-step process:

TABLE 8.8 Environmentally Responsible Process Assessments for Generic 1950s and 1990s Automobile Manufacturing*

Life Stage	Environmental Stressor					
	Materials Selection	Energy Use	Solid Residues	Liquid Residues	Gaseous Residues	Total
Resource provisioning	4	0	1	4	1	10/20
	4	2	2	4	3	15/20
Process implementation	0	0	2	2	2	6/20
	3	2	2	4	3	14/20
Primary process operation	0	1	1	1	1	4/20
	2	2	1	3	3	14/20
Complementary process operation	0	2	2	0	1	5/20
	3	3	3	2	3	14/20
Refurbishment, recycling, disposal	3	3	1	2	4	13/20
	4	1	3	4	4	16/20
Total	7/20	6/20	7/20	9/20	9/20	38/100
	16/20	10/20	11/20	17/20	16/20	70/100

*Upper numbers refer to 1950s automobile manufacturing processes, lower numbers to 1990s processes.

1. Review the process design and implementation plan with process suppliers and/or internal process designers, as well as with internal manufacturing engineers and supply-line management personnel.
2. Conduct a Design for Environment (DFE) review of the processes using the techniques discussed here or others found to be suitable.
3. Derive a list of recommendations from the results of the assessment, subdivided (if possible and appropriate) between those to be implemented by process design personnel, process operations personnel, and management.
4. Prioritize the recommendations.

It is important to emphasize the importance of step 4, because implementation of the results of a process SLCA assessment is generally enhanced if the recommendations that are generated in the report are prioritized. One way of doing so follows the approach of Chapter 4, where each recommendation is ranked on a "+/−" scale across the six considerations. Four of these are identical with those for products: technical feasibility, environmental sensitivity, matrix score improvement, and product management. Two differ slightly:

- *Economic benefit:* Rates the net cost of implementing a particular recommendation; "++" means the cost of implementing or operating the process would be significantly less if recommendation is incorporated. Here the total life-cycle cost should be considered. For example, some process components may cost more due to DFE constraints but yield a higher residual value when the equipment is retired.
- *CVA impact:* Accounts for the community-perceived value added by implementing a particular recommendation. In the case of products, DFE choices have rather

Section 8.5   Conducting the Assessment and Prioritizing the Recommendations   129

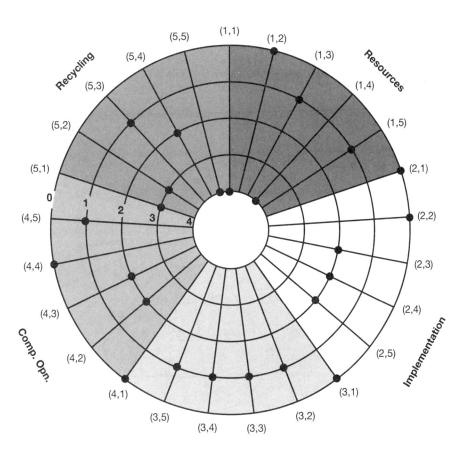

FIGURE 8.6   Comparative target plots for the display of the environmental impacts of the generic automobile manufacturing processes of the 1950s and of the 1990s.

small and indirect influences on the local community but large and direct effects on customers, so CVA refers to "customer value added." For processes, customers are subject only to small and indirect effects of DFE choices, but local communities can be affected substantially and directly. "++" means the DFE attribute has a very high perceived value on CVA. This parameter is completed from the perspective of the most environmentally conscious community organization.

To illustrate the technique for prioritizing process recommendations, Table 8.9 presents a list that might have been generated in connection with 1990 automotive manufacturing. Following the procedure outlined above, the recommendations are prioritized in Table 8.10, and displayed on a Pareto plot in Figure 8.7. This example diagram shows that, in the design area, three actions have high-priority ratings: (1) The elimination of plating processes, (2) specifying recycled content for process materials,

130 Chapter 8 Process Assessment by SLCA Matrix Approaches

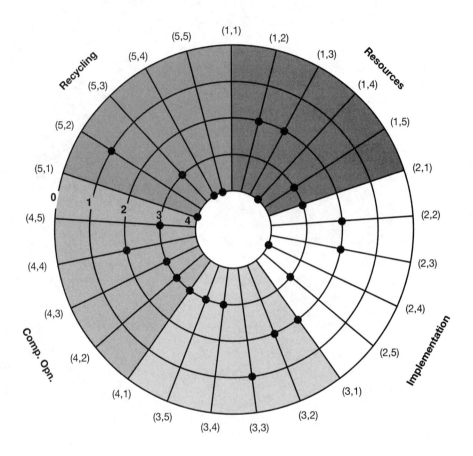

FIGURE 8.6 Continued.

and (3) marking of plastic parts. In the manufacturing area, two actions are identified for prompt attention: (1) Reducing the use of cutting and lubricating oils, and (2) recycling process solvents. For management, two actions merit priority treatment: (1) Incorporating DFE considerations into supplier purchasing agreements, and (2) encouraging or mandating recycled material in incoming chemical packaging.

It is important to note in closing this discussion of the environmental characteristics of industrial processes that the in-use process stage has been heavily regulated in the past decade or two while other life stages have been virtually ignored. The overriding need at the present time is to deal with the total life cycle: To work with suppliers and internal corporate engineers to design and implement processes that use materials wisely, that minimize environmental effects at process implementation, that avoid requiring undesirable complementary processes, and that have end-of-life

Section 8.5   Conducting the Assessment and Prioritizing the Recommendations   131

TABLE 8.9 Unprioritized Recommendations from the Generic Automobile Manufacturing Assessment

Recommended Activities for Design Organizations
- Eliminate the use of chromate as a metal preservative in favor of removable organic coatings.
- Develop efficient casting sand resue techniques.
- Eliminate all metal plating operations or sharply reduce residues therefrom.
- Specify >30% by weight recycled content for process materials.
- Have equipment suppliers mark all plastic parts, using the ISO standards.

Recommended Activities for Manusfacturing Organizations
- Reduce process energy consumption.
- When replacing or updating equipment, buy equipment designed for recycling.
- Reduce use of cutting and lubricating oils.
- Recycle all solvent used in processes.

Recommended Activities for Management Organizations
- Conduct a supplier environmental survey and develop a method of tracking supplier improvement.
- Rewrite specifications for processing chemical packaging material to encourage or mandate the use of some recycled material in their manufacture.
- Work with suppliers to develop take-back agreements on packaging entering the facility inc onnection with process components and process chemicals, cushioning, and exterior containers.
- Work with suppliers to minimize the diversity of packaging material entering the facility, so that recycling of solid waste may be optimized.
- Incorporate DFE considerations (DFE checklists and guidelines) in the performance specifications for all moduels and subsystems purchased from outside companies.

TABLE 8.10 A Prioritization Table for DFE Process Recommendations

Recommendation	TF	ES	ESI	EI	CVA	SI	Total
Design							
Eliminate chromate	+/−	++	+	+/−	+	−	3
Reuse casting sand	−	+/−	+	+	+	−−	0
Eliminate plating	++	++	++	+/−	++	+/−	8
Recycled content	++	+	+	+/−	+	+/−	5
Mark plastic parts	++	+	+	+/−	+	+/−	5
Operation							
Reduce energy use	+	+	+	+	+/−	+/−	4
Do DFR upgrades	+	+	+	+/−	+/−	+/−	3
Reduce oil use	++	+	+	+	+	−	5
Recycle solvents	++	+	+	+	+	−	5
Management							
Track suppliers	+	+	+	+/−	+	+/−	4
Recycled content	++	+	+	+/−	+	+/−	5
Packaging takeback	++	+	+	−	+	+/−	4
Packaging diversity	++	+	+	+/−	+/−	+/−	4
DFE purchasing	++	+	++	+/−	+	+/−	6

The column abbreviations are as follows: TF, technical feasibility; ES, environmental sensitivity; ESI, ERPS score improvement; EI, economic impact; CVA, community value added; SI, schedule impact. The method of scoring is described in Table 4.5.

designs that promote reuse and recycling of process equipment and chemicals, not their disposal.

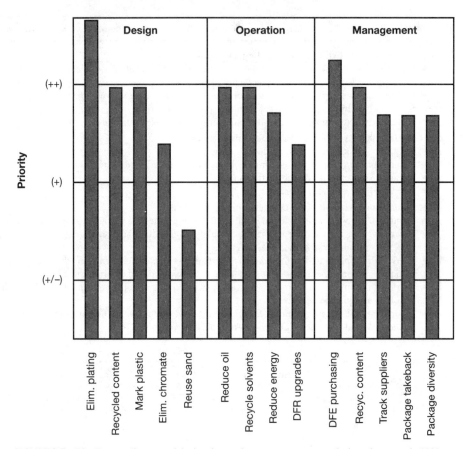

FIGURE 8.7 The Pareto diagram of design for environment recommendations for generic 1990 automotive manufacturing processes.

FURTHER READING

Allen, D.T., N. Bakshani, and K. Rosselot, *Pollution Prevention: Homework & Design Problems for Engineering Curricula* (New York: American Institute of Chemical Engineers, 1992).

Kalpakjian, S., *Manufacturing Engineering and Technology*, 2nd ed. (Reading, MA: Addison-Wesley, 1992).

EXERCISES

8.1. Choose a process with which you are familiar, such as fashioning a ceramic pot, constructing a small boat, or making a garment. Analyze the process using the approach of Section 8.3 and construct a process diagram like that shown in Figure 8.3.

8.2. Using the guidelines and protocols of Appendix B, evaluate, where possible, each matrix element (Table 8.1) for the process of Exercise 8.1. For those matrix elements where you could not perform an evaluation, what additional information was needed? How would you go about getting it?

8.3. Draw the target plot for the matrix completed in Exercise 8.2, omitting points where you were unable to assign a matrix element value. What life stages and what stressors are of most concern?

8.4. What recommendations would you make for improving the environmental attributes of the process chosen in Exercise 8.1? How would you prioritize the recommendations and what are the bases for the prioritization?

CHAPTER 9
Facility Assessment by SLCA Matrix Approaches

9.1 GOALS OF FACILITY ASSESSMENT

Just as products and processes can be made in environmentally responsible ways, so also can facilities be designed, built, operated, renovated, and recycled in an environmentally responsible manner. However, buildings have characteristics sufficiently different from products and processes so that facility assessment must be approached from a somewhat different framework. Among the obvious differences are that (1) the geographical location of a building has a strong influence on its design and construction (climate influences the degree to which heating and air conditioning are incorporated, for example), (2) it is common for the use of a building to change several times during its life span, (3) the end-of-life stage for a building is typically generations into the future, making it difficult to predict what materials recovery facilities may be desirable and what activities may be possible, and (4) very often, the use phase is predominant (e.g., 50 years of energy consumption is the controlling stressor).

Other factors affecting building life cycles are reminiscent of product assessments, but still present significant complications in the assessment process. For example, materials are used in different ways, some uses being more environmentally responsible than others (e.g., steel reinforcing bars in concrete present different problems for eventual recovery and recycling than do open steel beams over a manufacturing floor). Uses of a particular material may differ within a building as well, as in different window glazing on building exteriors directly exposed to sunlight. Finally, no matter what exemplary steps are taken in constructing the building, its overall environmental impact may be dominated by the way in which the facility is used, not by its structural features.

A number of architects, builders, and planners have approached the "green design" concept in recent years, and have established some recommended actions to guide thinking in the greening of facilities:

- Avoid building new roads or widening existing roads at the site.
- Minimize the "footprint" of construction operations.
- Minimize the use of heavy construction equipment.
- To the extent possible, preserve and/or restore local ecosystems.
- Reduce the quantity of materials used in construction.

- Select building materials that themselves have low environmental impacts.
- Design and operate energy-efficient facilities (i.e., efficient heating, cooling, lighting, machinery).
- Locate with complementary facilities that can utilize by-products.
- Reduce the use of water in the operation of buildings and grounds.
- Maximize the longevity and reuse of buildings.
- Recycle existing buildings and sites rather than developing undeveloped sites.
- Minimize materials waste during demolition and construction activities.

Designating a building as "green" generally involves evaluating the characteristics of the building's design and construction, especially materials selection, its infrastructure (lighting, heating, and so forth), and its eventual conversion to other uses or "deconstruction" and associated materials recycling.

All of these aspects of a building's relationship to the wider world are important, but the list is incomplete without the inclusion of the impacts of the activities carried on within the building during its useful life. For an industrial facility, for example, one needs to evaluate not only the structure itself but also the products that are manufactured within it, the processes that are used in that manufacture, and the ways in which other activities within the building are performed. Clearly a building cannot be truly green if the products, processes, and related operations associated with it are not.

9.2 LIFE STAGES OF INDUSTRIAL FACILITIES

Any assessment approach should ideally be applicable to all varieties of facilities. It is useful, therefore, to briefly review how facilities of various types can be approached from the life-cycle framework described in previous chapters. Table 9.1 lists a number of types of facilities and identifies their typical products and processes.

The first group of facilities are those that manufacture products for industrial or commercial customers. These are generally classical, "industrial" facilities, and have the

TABLE 9.1 Product and Process Characteristics of Typical Commercial and Industrial Facilities

Facility	Product	Process
Facilities Offering Products to Commercial Customers		
Ore smelter	Metal ingots	Smelting and refining of ores
Chemical works	Chemicals	Processing of chemical feedstocks
Appliance mfr.	Washing machines	Assemble products from components
Recycler	Components, materials	Disassemble and reprocess obsolete goods
Facilities Offering Products to Individual Consumers		
Hardware store	Tools, supplies	Unpacking, shelving
Grocery store	Food, related items	Unpacking, shelving, food processing
Garden center	Plants, related items	Agricultural activities
Restaurant	Meals	Food preparation

processing of materials or the production of components, subassemblies, or industrial infrastructure items such as machine tools as their focus. Assessments of this type of facility are reasonably straightforward.

The second group of facilities are those that exist for the purpose of offering tangible products directly to individual consumers. The simplest are those that perform the minimal level of processing: unpacking and shelving. Hardware stores, clothing stores, and small appliance stores are examples of this type. Grocery stores are similar, but typically perform some degree of food processing as well, such as meat cutting and packaging. Some facilities of this type have a more major involvement in processing, as with the agricultural activities of a typical garden center. Perhaps most like the industrial model is the restaurant, where materials (food) are transformed into products (meals) by specific processes (food preparation techniques).

Given the life-cycle perspective, the assessment of the environmental responsibility of a facility can be defined to include five stages or activities, as shown in Figure 9.1. Environmentally responsible facility (ERF) assessment need not and should not be applied only to manufacturing facilities, but rather to any facility engaged in any type of products or services: oil refineries, fast-food restaurants, residential structures, and so forth. The assessment will obviously be more complex in some cases than in others, but managers of facilities, no matter what the facility's function, should strive to environmentally responsible status.

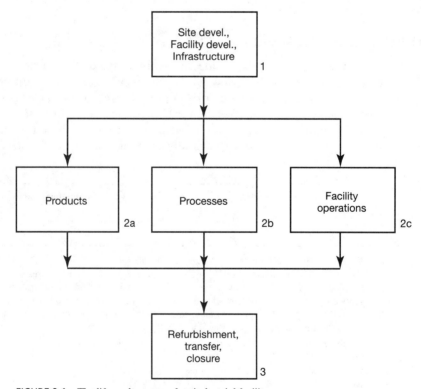

FIGURE 9.1 The life-cycle stages of an industrial facility.

Stage 1: Site Development, Facility Development, and Infrastructure

A significant factor in the degree of environmental responsibility of a facility is the site that is selected and the way in which that site is developed. If the facility is an extractive or materials processing operation (oil refining, ore smelting, and so on), the facility's geographical location will generally be constrained by the need to be near the resource. A manufacturing facility usually requires access to good transportation and a suitable work force, but may be otherwise unconstrained. Many other types of buildings can be located virtually anywhere.

Manufacturing plants have traditionally been in or near urban areas. Such locations often have suitable buildings available and have the advantages of drawing on a geographically concentrated work force and of using existing transportation and utility infrastructures. It may also be possible to add new operations to existing facilities, thereby avoiding many of the regulatory intricacies of establishing a completely new plant site. A promising recent development is the trend toward cooperative agreements between governments and industries for the reuse of these "brownfield" sites.

For facilities of any kind built on land previously undeveloped as industrial or commercial sites, ecological impacts on regional biodiversity may result, and we can anticipate added air emissions from new transportation and utility infrastructures. These effects can be minimized by working with existing infrastructures and developing the site with the maximum area left in natural form. Nonetheless, given the ready availability of commercial buildings and facilities in many cities and countries, such "greenfield" choices are hard to justify from an environmental perspective.

The construction of new facilities or the rehabilitation of existing facilities offers great opportunities for environmentally responsible action. The selection of materials, the methods of construction, and the handling of debris are all areas for attention. A substantial thrust is the use in new building construction of a variety of recycled materials, including wallboard of compressed paper, tiles from mining aggregate and ground glass, carpet pads of shredded tire rubber, and roof flashings of reclaimed copper and aluminum. In some instances, the value of wood that could be reclaimed from old buildings and reused made it economical to take the buildings apart rather than raze them and discard the rubble.

An activity that generally receives little attention is the disposal of construction debris. It has been estimated that as much as 25 percent by weight of all material brought to a building construction site is eventually consigned to the landfill, and approximately 20 percent of the waste flow to landfills is construction debris. It is probably difficult to avoid generating some of this debris—because of broken or defective materials, for example—but much of it could be reduced by an enhanced focus on manufacturing and building to standard dimensions, minimizing packaging, and promoting material reuse on site.

Stage 2a: Principal Business Activity: Products

Tangible products are items manufactured within the facility for sale to customers. Product assessment was discussed in Chapter 7.

Stage 2b: Principal Business Activity: Processes

Processes are the techniques, materials, and equipment used in the creation of products. Process assessment was discussed in Chapter 8.

Stage 2c: Facility Operations

The impact of any facility on the environment during its active life is often heavily weighted by transportation issues. As with many other aspects of industrial activities, tradeoffs are involved. For example, just-in-time delivery of components and modules has been hailed as a cost-effective and efficient boon for manufacturing. Nonetheless, it has been estimated that the largest contribution to the emissions that generate Tokyo smog comes from trucks making just-in-time deliveries. The corporations delivering and those receiving these components and modules should bear some degree of responsibility for those emissions. It is sometimes possible to reduce transport demands by improved scheduling and coordination, perhaps in concert with nearby industrial partners or by siting facilities near principal suppliers. Options may also exist for encouraging ride sharing, telecommuting, and other activities that reduce overall emissions from employee vehicles.

Material entering or leaving a facility also offers opportunities for useful action. To the extent that the material is related to products, it is captured by product SLCA assessments. Facilities receive and disperse much nonproduct material, however: food for employee cafeterias, office supplies, restroom supplies, and maintenance items such as lubricants, fertilizer, and road salt, to name just a few. An environmentally responsible facility should have a structured program to evaluate each incoming and outgoing materials stream and to tailor it, as well as its packaging, in environmentally responsible directions. Obviously, the most environmentally preferable products should be chosen to perform each function.

Facility energy use also requires careful scrutiny, as opportunities for improvement are always present. An example is industrial lighting systems, which are estimated to be responsible for between 5–10 percent of air pollution emissions overall. Another major energy consumer is the heating, ventilating, and air-conditioning systems. Office machines and computers in office buildings can also use significant amounts of energy. The environmental impacts chargeable to energy use generally occur elsewhere—emission of CO_2 from fossil fuel power plants, for example—but are no less real for not happening right at the facility.

As with many environmentally related business expenditures, energy costs for specific uses are often lumped in with "overhead" and not precisely known, yet the use of modern technology often has the potential to decrease energy expenditures by 50 percent or more, especially if energy use is charged to specific operations within the building and the managers of those operations are directed to monitor usage.

Stage 3: Facility Refurbishment, Transfer, and Closure

Just as environmentally responsible products are increasingly being designed for "product life extension," so should environmentally responsible facilities. Buildings and other structures contain substantial amounts of material with significant levels of embodied energy, and the (especially local) environmental disruption involved in the construction of new buildings and their related infrastructure is substantial. Clearly an environmentally responsible facility must be designed to be easily refurbished for new uses, to be transferred to new owners and operators with a minimum of alteration, and, if it must be closed, to permit recovery of materials, fixtures, and other components for reuse or recycling.

9.3 LIFE STAGES OF OFFICE AND RESIDENTIAL FACILITIES

The life-cycle approach outlined for industrial facilities needs to be modified for office or residential facilities, for which the concepts of "products" and "processes" are less applicable. Instead, the activities of the construction stage and the facility operation stage can be studied in greater detail. The result is five life stages defined as follows:

Stage 1a: Site and Infrastructure Development

This stage covers all aspects of the development of the site: ecological disturbance, provisioning of infrastructure, slope and drainage modifications, and the like.

Stage 1b: Facility Development

This stage treats all aspects of the construction of the building itself, including the choice of materials, their delivery to the site, techniques and equipment used in construction, and site cleanup.

Stage 2a: Facility Operations: Indoors

This stage includes activities taking place within the facility. It focuses on the use of energy and water, the choice and use of office and food supplies, choice and operation of heating, ventilation, and air conditioning equipment, and recycling or disposal of paper, food waste, and other debris.

Stage 2b: Facility Operations: Outdoors

This stage treats activities taking place outside the facility. It focuses on the use of energy and water, the approach to the maintenance of vegetation and plantings, snow plowing and salting, and other activities having a potential ecological impact.

Stage 3: Facility Refurbishment, Transfer, and Closure

This stage is identical with Stage 3 for industrial facilities.

The resulting life-stage diagram is shown in Figure 9.2.

9.4 ASSESSMENT APPROACHES FOR FACILITIES

Systematic valuation techniques for facilities are in the early stages of development. In one effort, a committee of the U.S. Green Buildings Council has produced an evalua-

140 Chapter 9 Facility Assessment by SLCA Matrix Approaches

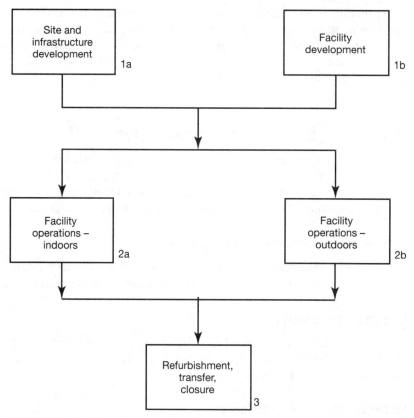

FIGURE 9.2 The life-cycle stages of an office or residential facility.

tion sheet that awards credits for the use of environmentally preferable building materials, ecologically sensitive landscaping and maintenance, energy efficiency, and water conservation, among other topics. Somewhat more elaborate schemes have been developed by the British Research Establishment and the University of British Columbia, where attempts are made to assess global, local, and indoor issues. The most ambitious project is by the U.S. National Institute of Standards and Technology, which has under development a comprehensive LCA tool for buildings.

While there are obvious benefits to be realized from the assessment tool development attempts mentioned above, there are some obvious inadequacies as well. The most significant is that certain life-cycle stages are emphasized while others are ignored; in particular, facility closure has not generally been addressed. In addition, most assessment tool development has been for residential housing, and is not easily adapted to commercial or industrial buildings. Finally, the emphasis tends to be on new construction, with little guidance available for the building rehabilitation activities that are potentially of great environmental advantage.

A suitable assessment system for environmentally responsible facilities should have the following characteristics: It should lend itself to direct comparisons among rated facilities, be usable and consistent across different assessment teams, encompass

TABLE 9.2 The Environmentally Responsible Industrial Facility Matrix

Facility life stage	Environmental Stressor				
	Ecological impacts	Energy use	Solid residues	Liquid residues	Gaseous residues
Site development, facility development, and infrastructure	1,1	1,2	1,3	1,4	1,5
Principal business activity—Products	2,1	2,2	2,3	2,4	2,5
Principal business activity—Processes	3,1	3,2	3,3	3,4	3,5
Facility operations	4,1	4,2	4,3	4,4	4,5
Refurbishment, transfer, and closure	5,1	5,2	5,3	5,4	5,5

(The numbers are the matrix element indices i, j.)

TABLE 9.3 The Environmentally Responsible Office and Residential Facility Matrix

Facility life stage	Environmental Stressor				
	Ecological impacts	Energy use	Solid residues	Liquid residues	Gaseous residues
Site and infrastructure development	1,1	1,2	1,3	1,4	1,5
Facility development	2,1	2,2	2,3	2,4	2,5
Facility operations—indoors	3,1	3,2	3,3	3,4	3,5
Facility operations—outdoors	4,1	4,2	4,3	4,4	4,5
Refurbishment, transfer, and closure	5,1	5,2	5,3	5,4	5,5

(The numbers are the matrix element indices i, j.)

all stages of facility life cycles and all relevant environmental stressors, and be simple enough to permit relatively quick and inexpensive assessments to be made. As with products and processes, it is feasible to develop the assessment system with a 5×5 matrix in which one dimension consists of life-cycle stages and activities and the other consists of environmental stressors. For the two types of facilities described above, those matrices are shown in Tables 9.2 and 9.3.

To preserve consistency with product and process assessments, the facilities analyst studies the characteristics of the facility and of the activities that occur within it, and assigns to each element of the matrix an integer rating from 0 (highest impact, a very negative evaluation) to 4 (lowest impact, an exemplary evaluation). She or he is guided in this task by experience, inspections of actual or planned facility characteristics, appropriate checklists (as in Appendix C), and other information. The process described here is purposely qualitative and utilitarian, and provides a numerical end point against which to measure improvement.

Once an evaluation has been made for each matrix element, the overall Environmentally Responsible Facility Rating is computed as the sum of the matrix element values:

$$R_{ERF} = \sum_i \sum_j M_{i,j} \qquad (9.1)$$

Because there are 25 matrix elements, a maximum facility rating is 100.

The assignment of a discrete value from zero to four for each matrix element implicitly assumes that the environmental impact implications of each element are equally important. An option for slightly increasing the complexity of the assessment (but perhaps increasing its utility as well) is to utilize detailed environmental impact information to apply weighting factors to the matrix elements. For example, a certain facility might be thought to generate most of its impacts as a consequence of the processes used within it and few related to facility operations, so the processes row could be weighted more heavily than before and the facility operations row weighted correspondingly lighter. Similarly, a judgment that global warming constituted more of a risk than did liquid residues might dictate an enhanced weighting of the energy use column and a corresponding decreased weighting of the liquid residue column. To the extent that an appropriate weighting scheme is obvious and non-contentious, its use will provide an improved perspective on the environmental burden of the facility being evaluated. Additional possibilities are discussed in Chapter 13.

The analyst needs to recognize that the life stages of Tables 9.2 and 9.3 are different from those for products or processes. For products, the five life stages that are considered are (1) premanufacture, (2) manufacture, (3) product delivery, (4) product use, and (5) recycling or disposal. For processes, the five life stages or activities are (1) premanufacture, (2) process implementation, (3) process operation, (4) complementary process operation, and (5) refurbishment, recycling, or disposal of the process equipment itself. For environmental stresses in the facilities assessment, four are the same as used for products and processes: energy consumption, solid residues, liquid residues, and gaseous residues.

Materials choice is replaced with biodiversity, because facility development, operation, and closure have such direct and important effects on local ecosystems.

In the industrial facility assessment approach, the first step is that products and processes related to the facility are identified and assessed by matrix approaches. The results are then incorporated into the assessment of the facility. Figure 9.3 shows the way in which the matrices are interrelated.

9.5 ASSESSING GENERIC AUTOMOBILE MANUFACTURING PLANTS

As a demonstration of the operation of the tools described above, it is instructive to perform environmentally responsible facility assessments on facilities more or less familiar to all, at least in concept: generic automobile manufacturing plants of the 1950s and 1990s. Some of the relevant characteristics of the facilities, their products, and their processes are outlined in Table 9.4. The products (automobiles) have already been assessed in Chapter 7; the processes by which they are made were discussed in Chapter 8.

While it is fairly straightforward to picture and describe typical automobiles of different eras and the processes by which they were made, there has historically been a wide variation in the types of manufacturing sites and their development. Two extremes are employed to illustrate assessment ranking differences. The site chosen to

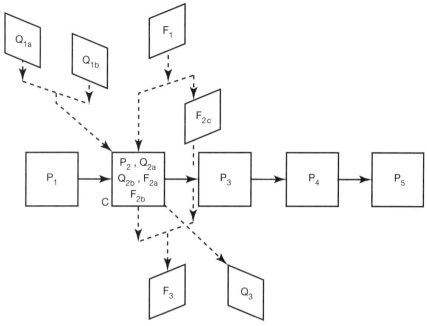

FIGURE 9.3 A schematic diagram of the interrelationships among product (P), process (Q), and facility (F) matrices. The common intersection area is designated C, and the subscripts refer to the life stages of Figs. 2.1, 8.2, and 9.1. The spectrum of environmental stressors applies to each life stage, so there is an additional dimension not seen in the diagram. The diagram shows a facility with one principal product produced by a single process, but the concept is readily generalized into multi-process, multi-product facilities.

represent the 1950s is an existing industrial site (a "brownfield site") with already available commercial power, road networks, and other municipal services. The building itself is of brick with steel framing and supports as required. Bus and trolley transportation are readily available to many of the employees. Heating is provided by coal burning, and air conditioning has not yet become routinely available. Incandescent lighting is abundant. Much of the land is grassed, mowed, and regularly treated with fertilizers and pesticides. The building itself is "purpose-built," as the British say: It is not designed with the idea of ever using it for any purpose other than automobile manufacture.

In contrast, the facility of the 1990s is built on a site new to commercial development, a "greenfield." Instead of modification or reconstruction of an existing building, natural land areas are developed together with the necessary infrastructure of roads, power lines, water and sewer services, and so forth. The structure is primarily concrete, but a wide variety of materials is used. Because the site is not near public transportation, private automobiles are used by employees to go to and from work. Heating is provided by natural gas, and modern air conditioning units with the potentially ozone-depleting refrigerant CFC-12 are used. In keeping with more enlightened modern practice, a substantial fraction of the grounds is maintained in a natural condition, and fertilizer and pesticide use is minimized.

TABLE 9.4 Salient Characteristics of Products, Processes, and Facilities for Generic Automobile Manufacturing Plants of the 1950s and 1990s

Characteristic	ca. 1950s	ca. 1990s
PRODUCT: The Automobile		
Material content (kg):		
Plastics	0	101
Aluminum	0	68
Metals	1583	1047
Rubber	85	61
Fluids	96	81
Other	137	76
Total Weight: (kg)	1901	1434
Fuel Efficiency (miles/gallon)	15	27
Exhaust Catalyst	No	Yes
Air Conditioning	CFC-12	CFC-134a
PROCESS: Auto Manufacture		
Energy use	Enormous	Substantial
Painting	Organic high-volatile	Aqueous low-volatile
Recycling	Some	Extensive
Process hardware		
Welding	Frequent	Ubiquitous
Conveyer belts	Numerous	Numerous
Complementary processes		
Metal cleaning	CFCs	Aqueous detergents
FACILITY: Auto Manufacturing Plant		
Site	"Brownfield"	"Greenfield"
Worker transport	Bus, trolley	Private auto
Heating	Coal	Natural gas
Lighting	Incandescent	High-eff. fluorescent
A/C Fluid	None	CFC-12
Grounds	Fertilizer, pesticides	Natural areas
Building mtls.	Brick, steel	Concrete, composites
Recycling	No	Yes
Closure	Demolition	Reuse-adaptable

The assessment is begun by treating the first life stage, that of facility site selection, development, and installation of infrastructure, guided by the matrix element checklists of Appendix C. The resulting ratings, and brief descriptions of the principal reasons for them, are shown in Table 9.5.

The second stage of facility assessment is that of the environmental responsibility of the products made within the facility (Table 9.6). For this stage, the Chapter 7 assessment of generic automobiles of the 1950s and 1990s will be used. For incorporation into the facilities assessment, the integer ratings for each of these impacts were summed over all five life stages, and then divided by five to put the rating on the same scale with the other facilities stages considered in this chapter. For example, the five materials choice ratings for the 1950s automobile were 2, 0, 3, 1, and 3. Their sum is 9, and the sum divided by 5 is 1.8, which is entered into Table 9.4.

The third activity stage is processes, for which the rating scores of Chapter 8 are utilized (Table 9.7).

Section 9.5 Assessing Generic Automobile Manufacturing Plants

TABLE 9.5 Site Selection, Development, and Infrastructure

Element Designation	Element Value and Explanation
1950s Plant	
Biodiversity (1,1)	2 (Brownfield site, but has biotic impacts.)
Energy use (1,2)	3 (Few modifications to energy infrastructure.)
Solid residue (1,3)	2 (Significant solid residues in site prep.)
Liq. residue (1,4)	3 (Modest liquid residues in site prep.)
Gas residue (1,5)	3 (Modest gaseous residues in site prep.)
1990s Plant	
Biodiversity (1,1)	1 (Greenfield site, large biotic impacts.)
Energy use (1,2)	0 (Complete new energy infrastructure.)
Solid residue (1,3)	1 (Abundant solid residues in site prep.)
Liq. residue (1,4)	3 (Modest liquid residues in site prep.)
Gas residue (1,5)	3 (Modest gaseous residues in site prep.)

TABLE 9.6 Principal Business Activity: Products

Element Designation	Element Value and Explanation
1950s Plant	
Matls. choice (2,1)	1.8 (CFC cleaning, virgin materials.)
Energy use (2,2)	1.4 (Fossil fuel energy use is very large.)
Solid residue (2,3)	2.2 (A number of components are difficult to recycle.)
Liq. residue (2,4)	2.6 (Fluid leakage during operation.)
Gas residue (2,5)	1.2 (No exhaust gas scrubbing; high emissions.)
1990s Plant	
Matls. choice (2,1)	2.6 (Petroleum is a resource in short supply.)
Energy use (2,2)	2.4 (Energy use during manufacture is fairly high.)
Solid residue (2,3)	2.8 (Modest residues of tires and obsolete parts.)
Liq. residue (2,4)	3.2 (Some liquid residues from cleaning and painting.)
Gas residue (2,5)	2.6 (CO_2, lead [sometimes] emissions.)

TABLE 9.7 Principal Business Activity: Processes

Element Designation	Element Value and Explanation
1950s Plant	
Materials choice (3,1)	2.0 (CFCs used for metal cleaning.)
Energy use (3,2)	1.8 (Substantial process energy use.)
Solid residues (3,3)	1.2 (No solid residue recycling.)
Liquid residues (3,4)	2.0 (Liquid residues from metal processing.)
Gaseous residues (3,5)	2.0 (No control of paint shop emissions.)
1990s Plant	
Materials choice (3,1)	2.6 (Few materials recoverable at facility demise.)
Energy use (3,2)	2.0 (High energy needs for process removal.)
Solid residues (3,3)	2.2 (Substantial solid residues at facility demise.)
Liquid residues (3,4)	2.8 (Moderate liquid residues.)
Gaseous residues (3,5)	3.0 (Few gaseous emissions concerns.)

TABLE 9.8 Facility Operations

Element Designation	Element Value and Explanation
1950s Plant	
Biodiversity (4,1)	0 (Pesticide use, all areas altered.)
Energy use (4,2)	1 (Large energy use in facility operations.)
Solid residue (4,3)	0 (No attempt to minimize solid residues.)
Liq. residue (4,4)	0 (Extensive liquid discharges.)
Gas residue (4,5)	2 (Moderate VOC emissions.)
1990s Plant	
Biodiversity (4,1)	3 (Natural areas, no pesticides.)
Energy use (4,2)	3 (Modest energy use in operations.)
Solid residue (4,3)	3 (Extensive waste minimization and recycling.)
Liq. residue (4,4)	3 (Extensive liquid residue treatment.)
Gas residue (4,5)	3 (Efficient gaseous residue controls.)

TABLE 9.9 Refurbishment/Transfer/Closure

Element Designation	Element Value and Explanation
1950s Plant	
Biodiversity (5,1)	1 (Major ecological impacts upon demolition.)
Energy use (5,2)	1 (Major energy use in demolition and clearing.)
Solid residue (5,3)	1 (Little reuse of materials possible.)
Liq. residue (5,4)	2 (Significant liquid residues when demolished.)
Gas residue (5,5)	2 (Significant gaseous residues when demolished.)
1990s Plant	
Biodiversity (5,1)	3 (Little ecological impact when reused.)
Energy use (5,2)	3 (Modest energy use when reused.)
Solid residue (5,3)	3 (Extensive reuse–low demolition probability.)
Liq. residue (5,4)	3 (Minor liquid residues when reused.)
Gas residue (5,5)	3 (Minor gaseous residues when reused.)

The fourth life stage, facility operations, has been described above; it encompasses any activities not directly related to products or processes. The ratings for the two generic facilities are shown in Table 9.8.

The final life stage assessment is for facility refurbishment, transfer, or closure (Table 9.9). In the case of the 1950s facility, the design employed no consideration of reuse, so total demolition constitutes the only reasonable option available. For the 1990s facility, which was designed so that interior walls may be modified, wires, cables, and pipes added readily and inexpensively, and building services updated in modular fashion, reuse is a reasonable expectation.

The completed matrices for the generic 1950s and 1990s automobile manufacturing plants are illustrated in Table 9.10. Examine first the values for the 1950s facility so far as life stages are concerned. The column at the far right of the table shows good environmental stewardship during site development, and moderate environmental stewardship regarding product design, manufacture, and use. The ratings for facility

TABLE 9.10 Environmentally Responsible Product Assessments for Generic 1950s and 1990s Automobile Manufacturing Plants*

Life Stage	Environmental Stressor					
	Biodiversity, Materials‡	Energy Use	Solid Residues	Liquid Residues	Gaseous Residues	Total
Site selection, development, infrastructure	2.0	3.0	2.0	3.0	3.0	13.0/20
	1.0	0.0	1.0	3.0	3.0	8.0/20
Principal business activity—Products	1.8	1.4	2.2	2.6	1.2	9.2/20
	2.6	2.4	2.8	3.2	2.6	13.6/20
Principal business activity—Processes	2.0	1.8	1.2	2.0	2.0	9.0/20
	2.6	2.0	2.2	2.8	3.0	12.6/20
Facility operations	0.0	1.0	0.0	0.0	2.0	3.0/20
	3.0	3.0	3.0	3.0	3.0	15.0/20
Refurbishment, transfer, and closure	1.0	1.0	1.0	2.0	2.0	7.0/20
	3.0	3.0	3.0	3.0	3.0	15.0/20
Total	6.8/20	8.2/20	6.4/20	9.6/20	10.2/20	41.2/100
	12.2/20	10.4/20	12.0/20	15.0/20	14.6/20	64.2/100

*Upper numbers refer to the 1950s facility, lower numbers to the 1990s facility.
‡The ratings in this column for life stages one, π four, and five refer to impacts on biodiversity, for stages two and three on choice of materials.

closure are poor, and those for facility operations are abysmal. On the basis of the environmental stressors shown by the column additions, the summed ratings for biodiversity and solid residues are particularly low. The overall rating of 41.2 is far below what might be desired.

In contrast, the overall rating for the 1990s facility is 64.2, much better than that of the earlier facility but still leaving plenty of room for improvement. The life stage that is particularly problematic is that of site selection and preparation. Energy use and solid residue generation receive the lowest ratings of the environmental stressors.

The target plots that complement Table 9.10 are shown in Figure 9.4. One instantly gets the sense of the areas in which improvement has occurred over the 20-year interval, and also where modern approaches appear deficient.

9.6 DISCUSSION

Corporations and individuals have traditionally been regarded as good citizens if they followed rules of behavior established by their societies. This reactive approach is now giving way to a proactive one. Processes were the first target of attention—emissions and energy use, for example. Products were next, as their environmentally related attributes were assessed and improved. Facilities are the third facet to receive attention. As techniques for facility assessment undergo further development, the construction and operation of green buildings will become an increasingly important aspect of environmental responsibility.

Two approaches to facility-environment interaction that are related to the discussion in this chapter are *green building design reviews* and *facility audits*. Their scope can

148 Chapter 9 Facility Assessment by SLCA Matrix Approaches

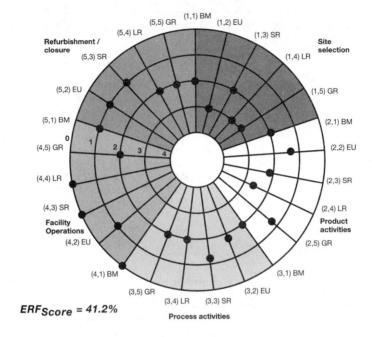

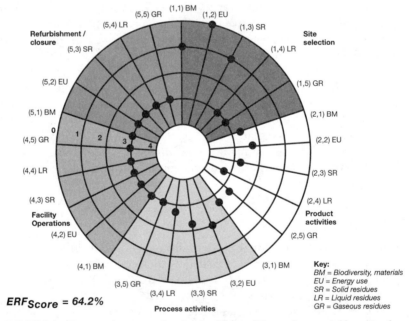

FIGURE 9.4 Comparative target plots for the display of the environmental impacts of generic automobile manufacturing facilities of the 1950s and of the 1990s.

be appreciated from the perspective of the life-cycle diagram shown in Figure 9.1. In green building design, the emphasis is on life stage 1 and the energy-related aspects of life stage 2c. In facility audits, the emphasis is on life stages 2a and 2b, with less attention given to life stage 2c. Green building design rarely considers life stage 3, and facility audits universally ignore the final life stage.

Utilizing green building design reviews and facility audits for buildings is reminiscent of the traditional regulatory division of environmental monitoring and control into air, water, and waste. In each approach, practitioners tend to think only about their turf, and the system as a whole seldom approaches optimization. Facility evaluation and the improvement of facility environmental performance is made difficult by the long time scales involved and by the different parties (builders, owners, renters, etc.) in control at different phases of the process, but it is clear that optimization can only be approached if the facility evaluation is carried out from a systems perspective.

FURTHER READING

Allenby, B.R. and T.E. Graedel, Defining the environmentally responsible facility, in *Proc. Third Nat. Acad. Eng. Workshop on Industrial Ecology* (Washington, D.C.: National Academy Press, 1997).

Graedel, T.E. and B.R. Allenby, Matrix approaches to green facility assessment, in *Second International Green Buildings Conference*, A.H. Fanney, K.M. Whitter, and T.B. Cohn, Eds., Spec. Pub. 888 (Gaithersburg, MD: National Institute of Standards and Technology, 1995) pp. 84–102.

Lippiatt, B.C. and G.A. Norris, Selecting environmentally and economically balanced building materials, in *Second International Green Buildings Conference*, A.H. Fanney, K.M. Whitter, and T.B. Cohn, Eds., Spec. Pub. 888 (Gaithersburg, MD: National Institute of Standards and Technology, 1995) pp. 37–46.

Seiter, D. and W.L. Doxsey, *Sustainable Building Sourcebook* (Austin, TX: Environmental and Conservation Services Dept., City of Austin, 1995).

U.S. Green Building Council, *Green Building Rating Systems* (Bethesda, MD, 1995).

EXERCISES

9.1. Choose a building with which you are familiar and to which you have free access. To the degree possible, comment on the compliance of that building with the green design guidelines of Section 9.1.

9.2. Using the guidelines and protocols of Appendix C, evaluate, where possible, each matrix element (Table 9.2) for the building of Exercise 9.1. For those matrix elements where you could not perform an evaluation, what additional information was needed? How would you go about getting it?

9.3. Draw the target plot for the matrix completed in Exercise 9.2, omitting points where you were unable to assign a matrix element value. What life stages and what stressors are of most concern?

9.4. What recommendations would you make for improving the environmental attributes of the building chosen in Exercise 9.1? How would you prioritize the recommendations and what are the bases for the prioritization?

CHAPTER 10
Services Assessment by SLCA Matrix Approaches

10.1 THE SERVICES INDUSTRY

The assessments discussed thus far have dealt with products of various kinds, with the processes used in their manufacture, and with the facilities in which product manufacture or sale takes place. In all cases, the actions generate an added value realized in material form and in the transfer of that material from one owner to another.

The service industry approaches customer services differently. The added value here is not centered on the transfer of material or finished products but on providing a function desired by the customer—a "dematerialized" added value, at least from the customer's standpoint. Perhaps the simplest service from a conceptual standpoint is a bank, which performs most functions electronically and provides only minimal levels of products: currency, paper confirmation of transactions, and the like. Schools have a somewhat greater diversity of materials, but again the product, education, is largely intangible. A hair salon is a different type of establishment: It employs a number of physical and chemical processes to perform a service on an item (hair) supplied by the customer. From the perspective of services assessment, a beauty salon, an appliance repair shop, and an auto body shop play the same role: maintenance of a customer-supplied item.

Low-technology services such as cooking and child care have, of course, existed since the dawn of civilization. Technologically related services have been with us a substantially shorter period of time; they include such things as providing rail transport from one city to another or furnishing intercontinental telegraph service. A more recent innovation is the practice of leasing equipment such as automobiles. In that case, a material product is transferred from supplier to customer, but only temporarily. What is being marketed in that case is the availability of individual transport, not the acquisition of several tons of steel, aluminum, plastic, glass, and the like. In the longer term, there have been proposals to structure automobile leases so that the lessee is provided with a vehicle wherever she or he happens to be: at home, on vacation in Jamaica, or on business in Greece, for example.

If industry will increasingly be providing services and not products, and there are many indications that such a situation will be the case, how can the environmental responsibility in such a business setting be evaluated? Who is responsible, and what assessment tools may be brought to bear? This is a little-explored topic, but one with a dynamic future. This chapter addresses these issues while recognizing that the existing answers to the questions are rather rudimentary, and that much evolution will doubtless occur.

10.2 TYPE ALPHA SERVICES: THE CUSTOMER COMES TO THE SERVICE

In the most common provisioning of services, called here *alpha services*, a service is provided in a fixed location and the customer travels to the service. An example is the dry cleaner, who receives clothing, treats it by one or more chemical and/or physical processes, and returns it. The product involved (the clothing) is owned by the customer, transferred temporarily to the service provider, and then recovered. From a DFE assessment standpoint, the impacts related to the building in which the service is provided, as well as the service itself, are chargeable to the provider. Depending on the philosophy of the assessor, the environmental impacts related to transportation could be chargeable either to the customer or to the provider. A number of examples of alpha services are given in Table 10.1.

The environmental responsibility of providing an alpha service is assessed much as one would perform a facility evaluation. Life stage 1 (Figure 10.1) treats either facility construction or the modifications required of the facility to make it suitable for the intended use. Stage 2 refers to the equipment needed to perform the service: computers, automotive repair hardware, and so forth. Stage 3a is similar to the manufacturing facility matrix, but refers to the DFE characteristics of the products being provided rather than those of products being made. Life stages 3b and 4 are evaluated as for manufacturing facilities.

10.3 TYPE BETA SERVICES: THE SERVICE GOES TO THE CUSTOMER

A second type of service is termed *beta*, and its distinguishing characteristic is that the provider performs the service at a customer location. Table 10.2 lists some examples, such as appliance repair in the customer's home. Providing and servicing leased photo-

TABLE 10.1 Product and Process Characteristics of Typical Commercial and Industrial Alpha Service Facilities

Facility	Product	Process
Auto repair	Reconditioned auto	Part and function maintenance
Bank	Financial services	In-person transactions
Lawyer's office	Legal services	Consultation, document preparation
Retail stores	Various	Consumer sales
Hospital	Health maintenance	Medical care
Dry cleaner	Clean clothing	Solvent cleaning
School	Education	Classroom instruction
Hair salon	Hair maintenance	Chemical and physical treatments
Hotel	Rental rooms	Clean, maintain, and provide

152 Chapter 10 Services Assessment by SLCA Matrix Approaches

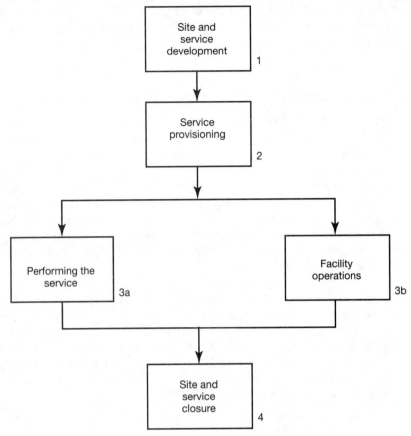

FIGURE 10.1 The life stages of a service.

TABLE 10.2 Product and Process Characteristics of Typical Commercial and Industrial Beta Service Facilities

Facility	Product	Process
Postal delivery	Mail handling	Mail sorting and delivery
Product rental or leasing	As rented or leased	Temporary product transfer
Appliance repair	Reconditioned appliance	Part and function maintenance
Building/remodeling	Renovated structure	Carpentry
Grounds care	Property maintenance	Mowing, fertilizing, etc.
Package handling	Transport of packages	Pickup, movement, delivery
Burglar alarms—installation	Installed alarm	Carpentry
Water, gas, sewer utilities	Deliver/receive flow streams	Install pipes, process flows

copy machines is a common commercial example. Another type of beta service is the leasing of a television set. In this case, the material is owned by the service provider, transferred temporarily to the user, and then recovered. From the DFE analyst's standpoint, the inputs related to providing the service, including transportation, are chargeable to the provider, while those related to the facility refer to the facility from which the service originates.

The environmental responsibility of providing a beta service is assessed similarly to alpha services because beta services require a location out of which to do business. The only principal distinction is that stage 3a, performing the service, is accomplished at the customer's location and thus the environmental impacts related to transportation (which in some cases can be significant) must be included in the Stage 3a assessment.

10.4 TYPE GAMMA SERVICES: REMOTE PROVISIONING

The advent of modern technology has enabled a new class of services to arise in recent years. Called *gamma services*, these are provided without either the customer traveling to the service or the service traveling to the customer. Rather, the service is provided by electronic means, such as bank services by telephone or other examples given in Table 10.3. In its more extensive implementations, gamma services permit entire careers to be accomplished with a minimum of physical movement through the use of computers, telephones, fax machines, modems, and associated software and hardware.

As with other service activities, one wants to know how to optimize gamma services from an environmental perspective. A characteristic of gamma services is that even though there is no direct physical contact with the customer, accomplishing the service inevitably requires that hardware and supporting personnel be located within a facility of the corporation supplying the services. For the gamma assessment, therefore, three life stages relate to the facility itself:

- Life stage 1—site selection, development, and infrastructure
- Life stage 3b—facility operations
- Life stage 4—refurbishment, transfer, closure

The other two life stages refer to the equipment used to provide the service: Life stage 2—equipment DFE characteristics; and life stage 3a—equipment operation. In addition, stage 3a includes the environmental implications of customer interactions such as energy use, customer service representatives, and so forth.

TABLE 10.3 Product and Process Characteristics of Typical Commercial and Industrial Gamma Service Facilities

Facility	Product	Process
Bank	Financial services	Electronic transactions
Burglar alarms—operation	Building monitoring	Electronic communication
Telecommuting	Business information	Electronic communication

10.5 THE SERVICES ASSESSMENT MATRIX

As with products, processes, and facilities, services can be evaluated by SLCA matrix techniques. In the adaptation of the 5×5 product matrix of Chapter 7, shown in Table 10.4, the environmental stressors are retained but the life stages are those described below.

Stage 1: Site and Service Development

To develop and deliver a service, either a service operations facility or an office facility is needed. Significant factors in the degree of environmental responsibility of a facility are the site selection, the way in which the site is developed, and the infrastructure of the site. Most service facilities must, of necessity, be located near customers who will require transportation in some form. (In many suburban areas, the private automobile is the only reasonable alternative.) The site selection can influence this transportation need. Offices can be situated anywhere as long as the commute for employees is reasonable. And, given the new telecommunications infrastructures with capabilities such as forwarding telephone calls, answering services, and computer network capabilities with e-mail and online data exchange, the need for being in the same physical location is reduced. The question then becomes whether the infrastructure is supporting the optimization in space and transportation.

The impact of the service provider on the environmental impact of the facility varies. Often the service provider is leasing the facility. In that case, the selection and contract phase is the only opportunity to influence the facility's inherent environmental impact. The facilities matrix (Chapter 9) can be used to negotiate contractual arrangements to optimize the facility's environmental impact. In the more flexible case where the service provider is selecting and building its own facility, the facilities matrix can be used in the planning and design stage as well as during facility operation.

Stage 2: Service Provisioning

To develop and operate a service, it is generally necessary to purchase or lease products such as desktop computers, office machines, maintenance equipment and supplies, vehicles, and so forth. The specification of the environmental characteristics of these items is the responsibility of the service provider, and should be carried out as the

TABLE 10.4 The Environmentally Responsible Services Matrix

Services Activity	Environmental stressor				
	Biodiversity/ materials	Energy use	Solid residues	Liquid residues	Gaseous residues
Site and service development	1,1	1,2	1,3	1,4	1,5
Service provisioning	2,1	2,2	2,3	2,4	2,5
Facility operations	3,1	3,2	3,3	3,4	3,5
Providing the service	4,1	4,2	4,3	4,4	4,5
Site and service closure	5,1	5,2	5,3	5,4	5,5

(The numbers are the matrix element indices i, j.)

equipment is purchased or leased. Considerations include modularity, maintainability, in-use resource consumption, and design for recycling.

Stage 3a: Performing the Service

This stage deals with the actual operation of performing the service. Because services cover such a broad spectrum of activities, it is difficult to provide general advice on environmentally responsible actions except to emphasize that all relevant environmental stressors must be addressed: biotic and toxic effects, energy use, and the generation and disposition of residues. For example, providers of grounds maintenance services need to carefully consider how fertilizers and pesticides are used, while beauty salons need to select treatment products carefully. Energy consumption should be a consideration for all service providers, not only in their facilities but also in the equipment that they use to provide the service. If there are service provider vehicles, they should be energy-efficient and their routes optimized.

Residues of all kinds should be examined with the idea of minimization or cessation. Hospitals need to decrease the generation of medically related solid waste, for example, and dry cleaners need to control cleaning fluid emission. Packaging of all kinds is generally susceptible to reduction if service providers work with their suppliers on optimizing packaging materials and approaches.

Stage 3b: Facility Operations

The concerns of this services life stage are identical with those for stage 2c of the facilities assessment of Chapter 9.

Stage 4: Site and Services Closure

The end of the life of a service will vary from the end of the use of an idea or the deletion of a file in computer memory to the recycling of products, the closure of a process, or the reuse of the facility. Ending a telephone service, for example, due to a customer's moving to another location or to her use of an alternate cable or cellular service, is an administrative process involving a little computer reprogramming and a few minutes work at the connections panel in a telecommunications equipment building. The freed telephone capacity can then be used for other customers. The service cable probably would be left in place so that service could be easily restarted if desired. The telephone that was the instrument that provided the service could, if leased, be returned to the servicing company. If the phone is owned by the user, the process is less simple. It might be stored, sold during a garage sale, or discarded in municipal solid waste receptacles. Government policies have thus far not dealt very well with this situation. In Germany, for example, the producer of the telephone has a take-back obligation, but the user does not have a give-back obligation.

More comprehensive service closure involves not only the service itself, but also the site and structures connected with the service. Just as environmentally responsible products are increasingly being designed for product life extension and reuse, so should the buildings connected with services. These structures contain substantial amounts of material with significant levels of embedded energy (i.e., the energy needed to extract, process, and emplace the material), and the (especially local) environmental disruption involved in the construction of new buildings and related infrastructure is substantial. Thus, an environmentally responsible service facility should be designed to be easily refurbished for new uses, to be transferred to new owners and operators with a minimum of alteration, and, if it must be closed, to permit recovery and reuse of obsolete stock, unused materials, fixtures, and other components. Failing that, both the building and its contents should be designed and managed so as to allow for efficient recycling.

10.6 CONTRACTS FOR BUILDINGS AND EQUIPMENT

It is common for manufacturing facilities to be sufficiently specialized that they are under the direct ownership and control of the operating corporation. It is equally common for service facilities and facilities management to be leased. This situation brings with it the potential for environmental responsibility to fall between the cracks: The lessor has little incentive to build, operate, and close the building in an environmentally responsible manner and the lessee has minimal and transitory responsibility for the facility.

If a lessee takes space in an existing building, the site selection, development, and infrastructure phase has already occurred, and the end-of-life characteristics of the building are established. In the lease negotiations, however, the lessee can attempt to influence the facility operations stage—property maintenance, food service operations, and the like. Initial tenants, especially large ones, will obviously have more bargaining power than subsequent tenants, especially small ones, but changes are often possible. And, in any case, the service operations within the facilities are under the control of the tenant.

Equipment and supplies, that is, service provisioning, is an area in which environmental performance can be required in supplier contracts. Cabinetry, flooring, business machines—all should be selected from environmentally responsible suppliers manufacturing environmentally responsible products. Pre-owned equipment such as office furniture, restaurant supplies, and electronic equipment is often available, and is an environmentally sound choice if otherwise suitable.

10.7 ASSESSMENTS OF GENERIC SERVICES

Service industries perform tasks that otherwise would be performed by the customer or the customer's organization. As an example, consider the customer who wants to replace a leaky fuel pump on his automobile with a fully operational fuel pump. Imagine that he (for convenience in this description a male is assumed) is to do this repair job himself. His activities are indicated in Figure 10.2.

Our handyman has previously equipped himself with appropriate tools and equipment in his original "service provisioning" operation. To do so, he drove to the hardware store, purchased the items, returned home, and put the items in a convenient storage location. The shopping trip generated tailpipe emissions from his automobile.

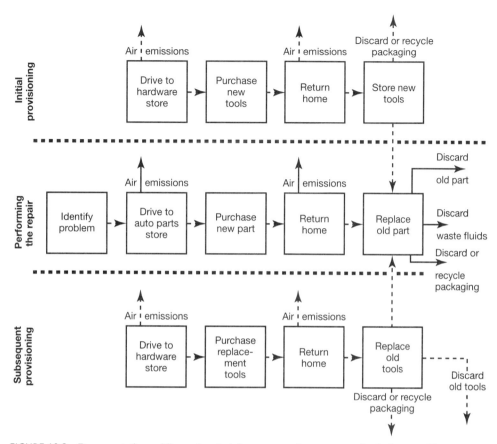

FIGURE 10.2 Representations of flows of materials, energy, and components for do-it-yourself automotive repair. Solid lines link steps in providing the service, dashed lines link steps in provisioning for the service.

He may or may not have recycled the packaging material from the new tools and equipment.

When he discovered the fuel leak, the handyman drove his automobile onto a ramp, scooted under on a wheeled "dolly," and confirmed that the fuel pump needed to be replaced. He then got into his other vehicle, drove to the auto parts store, purchased the appropriate replacement pump, and returned home. Scooting back under his vehicle, he exchanged the defective fuel pump for its newly acquired replacement, and finished by discarding the old pump, any fluids that had leaked during the repair, and the packaging from the new pump.

A few years later, the metal ramp and some of the other equipment used for automotive maintenance became unserviceable due to rust, or became obsolete because of the development of improved equipment. This required an updated provisioning, in which one or more trips to the hardware or auto parts store were again involved. The old equipment was probably discarded rather than recycled.

What are some of the environmental implications of the home repair operation? An obvious one is that a lot of driving was involved in order to prepare to perform the repair. A second is that a person performing an occasional repair has little incentive to purchase environmentally responsible equipment or to dispose of obsolete or defective equipment or parts in a responsible manner. Although recycling from the home has become common, the system is set up to handle items that are recycled in substantial volume: newspapers, bottles, cans, and the like. Old fuel pumps and servicing ramps may or may not be collected; even if they are, the chances that they will find their way into an appropriate refurbishment or recycling stream is small. The incentive to recycle waste fluids, such as contaminated oil or lubricants, is also small because it is often difficult for an individual to locate a facility willing to accept such materials. A home handyman who performs a large number of repairs may indeed locate appropriate recycling sources, but such an operation is more like a service business than a handyman operation. At some point, the individual chooses to let others repair the vehicles, and old equipment, excess solvents, and spare parts are discarded.

The alternative to home repair of a vehicle is repair by an automobile repair service. Consider the steps in replacing a fuel pump if this second option is chosen (Figure 10.3). The service is housed in a building built for the purpose and supplied with appropriate equipment. The initial provisioning stage is accomplished largely by an equipment supplier, who delivers and installs the equipment and (more and more commonly) takes away the packaging. Periodically thereafter, other suppliers bring new and rebuilt parts directly to the facility. When a customer arrives, the problem is diagnosed, the appropriate part or parts are retrieved from storage, the repair is accomplished, and the vehicle is returned to the customer. As with the home handyman, equipment must be periodically replaced. The new equipment is generally provided by suppliers who deliver and install the new items, recycle the old, and deal with any packaging and installation residues.

In the case of home automobile repair, building-related considerations are unimportant from the standpoint of performing an environmental assessment of the repair process because residential construction, operation, and closure are not influenced by the occasional automobile repair operation. The repair service, on the other hand, generally resides in a building built or modified strictly for its purpose. Its non-service functions (such as grounds maintenance) are probably also under the direct control of the service provider. Consequently, several life stages of the service can be assessed by facility assessment approaches using techniques similar to those used to look at manufactured products. So far as providing the service is concerned, several aspects tend to demonstrate environmental benefits. First, dealing with supplies in bulk minimizes delivery and associated air pollution emissions. Second, equipment suppliers can be chosen for the environmental responsibility of their products and installation procedures. Third, the flow of significant amounts of defective but rebuildable items such as pumps and alternators encourages their transfer to efficient refurbishing contractors. Finally, the regularity with which automotive fluids and part packaging materials are accumulated makes their recycling relatively efficient.

Can the matrix technique evaluate this alpha service? As before, one can use a 5×5 matrix, as shown in Table 10.4. For comparative purposes, it is instructive to compare the home automobile repair approach with a contemporary, commercial, environmentally responsible automobile repair facility and an environmentally irresponsible

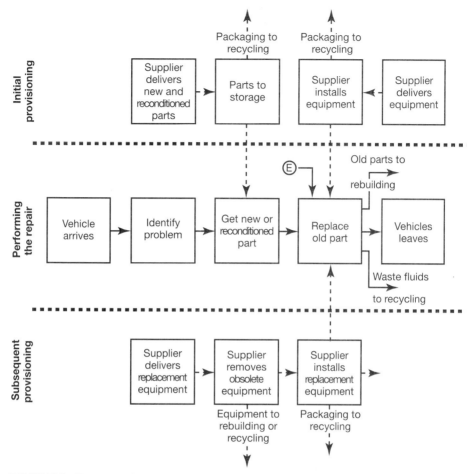

FIGURE 10.3 Representations of flows of materials, energy, and components for an automotive repair service. Solid lines link steps in providing the service, dashed lines link steps in provisioning for the service. A circled E indicates the use of a significant amount of energy.

one. The functional unit being considered is something like "providing a service facility and performing a typical service function." The characteristics of these three approaches are shown in Table 10.5. The home repair approach uses existing garage facilities, with the principal modifications being to add extra heating and lighting to make repairs possible during the winter months.

The irresponsible service is envisioned as taking place in a facility developed on a previously undisturbed site, far from public transportation or other infrastructure. (For purposes of this illustration, it is assumed that proximity to customers is roughly equivalent.) The service provider makes no attempt to deal with suppliers of environmentally responsible equipment. The facility is operated in an environmentally unsound manner, and during the auto repair activities no parts recycling is practiced and chlorinated solvents are used for metal cleaning. Should the facility eventually be abandoned, it must be demolished, because it is not designed for reuse.

TABLE 10.5 Salient Characteristics of Generic Environmentally Responsible and Environmentally Irresponsible Auto Repair Shops

Characteristic	Home repair	Irresponsible	Responsible
Facility Development			
Site	Residential	"Greenfield"	"Brownfield"
Worker transport	Not applicable	Private auto	Bus, trolley
Building mtls.	Wood frame	Brick, steel	Concrete, composites
Facility provisioning	Existing	New construction	Renovation
Service Provisioning			
"Green" equipment	No	No	Yes
"Green" installation	No	No	Yes
Performing the Service			
Energy use	High	Modest	Modest
Metal cleaning	Light oils	Chlorinated solvents	Aqueous detergents
Recycling of packaging	No	No	Yes
Recycling of defective parts	No	No	Yes
Recycling of fluids	No	No	Yes
Facility Operations			
Heating	Oil	Coal	Natural gas
Lighting	Incandescent	Incandescent	High-eff. fluorescent
A/C fluid	None	CFC-12	HFC-134a
Grounds	Not applicable	Fertilizer, pesticides	Natural areas
Recycling	Not applicable	No	Yes
Site and Service Closure			
Solvents recycled	No	No	Yes
Hardware recycled	No	No	Yes
Closure	Site remains	Demolition	Reuse-adaptable

The third option, the environmentally responsible service, is envisioned as being performed within a modified existing building at a site near public transportation and other infrastructure. Environmentally conscious suppliers are used, the facility is operated in a responsible manner, and full recycling is practiced. Should the service end, the building can be reused.

The first life stage, site and service development, demonstrates the capability of the matrix system to distinguish between these two approaches to providing auto repair services. The ratings are shown in Table 10.6.

Life stage 2 of services assessment is that of services provisioning. This stage evaluates the performance of suppliers in building and installing equipment. Automobile repair equipment tends to be fairly benign environmentally, and differences among the environmental performance of suppliers are not marked. The ratings are shown in Table 10.7.

Life stage 3a is performing the service. In any service that involves replacing defective or obsolete parts and components with others, packaging and disposal will be major issues. Emissions come into play as well. The assigned ratings are shown in Table 10.8.

Life stage 3b, facility operations, has been described above; it encompasses any activities not directly related to products or processes. The ratings for the service operations are shown in Table 10.9.

TABLE 10.6   Site and Service Development Ratings

Element Designation	Element Value and Explanation
HR Approach	
Ecol. Impacts (1,1)	4 (No additional site development.)
Energy use (1,2)	4 (No additional site development.)
Solid residue (1,3)	4 (No additional site development.)
Liq. residue (1,4)	4 (No additional site development.)
Gas residue (1,5)	4 (No additional site development.)
EI Facility	
Ecol. impacts (1,1)	1 (Greenfield site, large biotic impacts.)
Energy use (1,2)	0 (Completely new energy infrastructure.)
Solid residue (1,3)	2 (Significant solid residues in site prep.)
Liq. residue (1,4)	2 (Significant liquid residues in site prep.)
Gas residue (1,5)	3 (Modest gaseous residues in site prep.)
ER Facility	
Ecol. impacts (1,1)	3 (Brownfield site, small biotic impacts.)
Energy use (1,2)	3 (Minor modifications to existing infrastructure.)
Solid residue (1,3)	3 (Modest solid residues in site prep.)
Liq. residue (1,4)	3 (Modest liquid residues in site prep.)
Gas residue (1,5)	3 (Modest gaseous residues in site prep.)

TABLE 10.7   Service Provisioning Ratings

Element Designation	Element Value and Explanation
HR Approach	
Ecol. Impacts (2,1)	3 (Some equipment is chrome plated.)
Energy use (2,2)	4 (No additional site development.)
Solid residue (2,3)	2 (Equipment packaging residues.)
Liq. residue (2,4)	4 (No concerns.)
Gas residue (2,5)	4 (No concerns.)
EI Facility	
Matls. choice (2,1)	3 (Some equipment is chrome plated.)
Energy use (2,2)	3 (Modest energy use during installation.)
Solid residue (2,3)	1 (Substantial equipment packaging residues.)
Liq. residue (2,4)	4 (No concerns.)
Gas residue (2,5)	4 (No concerns.)
ER Facility	
Matls. choice (2,1)	4 (No concerns.)
Energy use (2,2)	3 (Modest energy use during installation.)
Solid residue (2,3)	3 (Modest residues from installation.)
Liq. residue (2,4)	4 (No concerns.)
Gas residue (2,5)	4 (No concerns.)

The final life stage assessment is for site and service refurbishment, transfer, or closure (Table 10.10). The difference between the "designed for disassembly and reuse" facility and the traditional facility is evident in the scoring.

TABLE 10.8 Performing the Service Ratings

Element Designation	Element Value and Explanation
HR Approach	
Ecol. Impacts (3,1)	2 (Light oils used for metal cleaning.)
Energy use (3,2)	1 (Extensive travel required.)
Solid residue (3,3)	0 (No solid residue recycling.)
Liq. residue (3,4)	0 (No liquid residue recycling.)
Gas residue (3,5)	1 (Substantial tailpipe emissions.)
EI Facility	
Materials choice (3,1)	1 (Chlorinated solvents used for metal cleaning.)
Energy use (3,2)	3 (Modest energy use.)
Solid residues (3,3)	0 (No solid residue recycling.)
Liquid residues (3,4)	0 (No liquid residue recycling.)
Gaseous residues (3,5)	2 (No control of vapor emissions.)
ER Facility	
Materials choice (3,1)	3 (Few concerns.)
Energy use (3,2)	3 (Modest energy use.)
Solid residues (3,3)	3 (Solid residue recycling.)
Liquid residues (3,4)	3 (Liquid residue recycling.)
Gaseous residues (3,5)	3 (Few gaseous emissions concerns.)

TABLE 10.9 Facility Operations Ratings

Element Designation	Element Value and Explanation
HR Approach	
Ecol. Impacts (4,1)	4 (No concerns.)
Energy use (4,2)	3 (Modest energy use.)
Solid residue (4,3)	4 (No concerns.)
Liq. residue (4,4)	4 (No concerns.)
Gas residue (4,5)	3 (Some volatile organic emissions.)
EI Facility	
Ecol. impacts (4,1)	0 (Pesticide use, all areas altered.)
Energy use (4,2)	1 (Large energy use in facility operations.)
Solid residue (4,3)	0 (No attempt to minimize solid residues.)
Liq. residue (4,4)	0 (Extensive liquid discharges.)
Gas residue (4,5)	2 (Moderate VOC emissions.)
ER Facility	
Ecol. impacts (4,1)	3 (Natural areas, no pesticides.)
Energy use (4,2)	3 (Modest energy use in operations.)
Solid residue (4,3)	3 (Extensive waste minimization and recycling.)
Liq. residue (4,4)	3 (Extensive liquid residue treatment.)
Gas residue (4,5)	3 (Efficient gaseous residue controls.)

The completed matrices for the auto repair approaches are illustrated in Table 10.11. Examine first the values for the irresponsible facility so far as life stages are concerned. The column at the far right of the table shows moderate environmental stew-

Section 10.7 Assessments of Generic Services

TABLE 10.10 Refurbishment/Transfer/Closure Ratings

Element Designation	Element Value and Explanation
HR Approach	
Ecol. Impacts (5,1)	4 (No concerns.)
Energy use (5,2)	4 (No concerns.)
Solid residue (5,3)	2 (Hardware discarded.)
Liq. residue (5,4)	1 (Solvents discarded.)
Gas residue (5,5)	4 (No concerns.)
EI Facility	
Ecol. impacts (5,1)	1 (Major ecological impacts upon demolition.)
Energy use (5,2)	1 (Major energy use in demolition and clearing.)
Solid residue (5,3)	1 (Little reuse of materials possible.)
Liq. residue (5,4)	2 (Significant liquid residues when demolished.)
Gas residue (5,5)	2 (Significant gaseous residues when demolished.)
ER Facility	
Ecol. impacts (5,1)	3 (Little ecological impact when facility reused.)
Energy use (5,2)	3 (Modest energy use when facility reused.)
Solid residue (5,3)	3 (Extensive reuse–low demolition probability.)
Liq. residue (5,4)	3 (Minor liquid residues when facility reused.)
Gas residue (5,5)	3 (Minor gaseous residues when facility reused.)

TABLE 10.11 Environmentally Responsible Service Assessments for Generic Automobile Repair Shops[*]

	Environmental Stressor					
Life stage	Biodiversity/ materials	Energy use	Solid residues	Liquid residues	Gaseous residues	Total
Site and services development	4	4	4	4	4	20/20
	1	0	2	2	3	8/20
	3	3	3	3	3	15/20
Service provisioning	3	4	2	4	4	17/20
	3	3	1	4	4	15/20
	4	3	3	4	4	18/20
Performing the service	2	1	0	0	1	4/20
	1	3	0	0	2	6/20
	3	3	3	3	3	15/20
Facility operations	4	3	4	4	3	18/20
	0	1	0	0	2	3/20
	3	3	3	3	3	15/20
Site and service closure	4	4	2	1	4	6/20
	1	1	1	2	2	7/20
	3	3	3	3	3	15/20
Total	17/20	16/20	12/20	13/20	16/20	74/100
	6/20	8/20	4/20	8/20	13/20	39/100
	16/20	15/20	15/20	16/20	16/20	78/100

[*]Upper numbers refer to home automobile repair, middle numbers to an environmentally irresponsible commercial auto repair service, and the lower numbers to an environmentally responsible one.

ardship during service provisioning. The ratings during site and service development and closure are poor, and for facility operations and performing the service the ratings are abysmal. The overall rating of 39 is far below what might be desired. In contrast, the overall rating for the responsible facility is 78, with all life stages being relatively satisfactory, although still with plenty of room for improvement. The score for home repair is nearly as good, deficiencies in providing the service (largely due to extensive transportation) being partially compensated by the use with little additional impact of an existing facility.

The target plots for the three approaches are shown in Figure 10.4. Here the differences in life stages 3a and 3b between the two commercial services are particularly obvious. The home repair approach does well in life stages related to the facility,

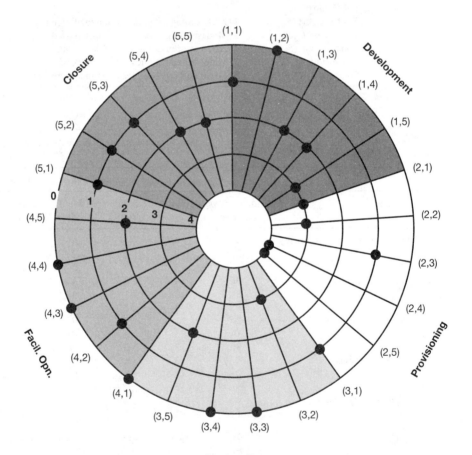

FIGURE 10.4 Comparative target plots for the display of the environmental impacts of two automobile repair shops, one environmentally irresponsible (a), the other environmentally responsible (b).

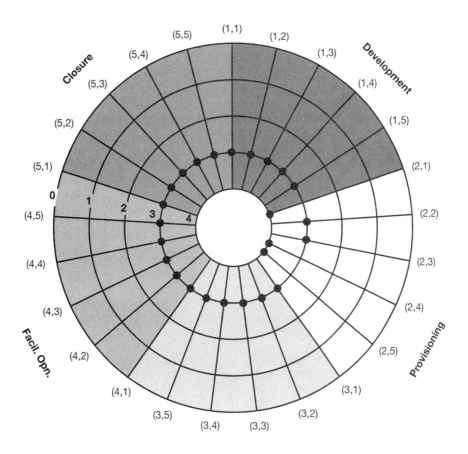

FIGURE 10.4 Continued.

because it is already existing and the repair operations cause only small impacts, but this approach is especially deficient in life stage 3a, the rating being heavily impacted by extensive transportation needs and lack of recycling.

As a second example of a service assessment, consider two banks, one of which operates in the traditional way in which the customer conducts transactions at the bank (an alpha service), the other in which all transactions are accomplished electronically (a gamma service). From the customer's standpoint, the event sequences connected with a simple action, such as transferring funds from a savings account to a checking account, are shown in Figure 10.5a. The environmentally preferable service is clearly the gamma service, which avoids the emission of significant amounts of automotive air pollutants. Some electrical energy is used in either mode, but the amount is small.

Now look at the bank transaction from the standpoint of the service provider. For the alpha service, the bank must provide both "front-office" equipment (counters,

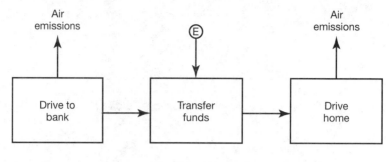

The Alpha Service

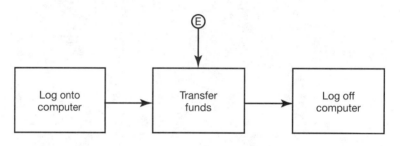

The Gamma Service

FIGURE 10.5   Representations of flows of materials and resources for two methods of banking. (a) Alpha and gamma banking from the customer's perspective (diagram above); (b) alpha banking from the bank's perspective (diagram on p. 167); (c) gamma banking from the bank's perspective (diagram on p. 168).

desks, teller windows, etc.) and "back-office" equipment (computers, optical character readers, communications links, etc.). The provisioning must be accomplished both initially and periodically. The facility must support both front-office and back-office functions.

A bank providing a gamma service retains the requirement of providing a suitable building and of operating and closing it in an environmentally responsible manner, but it no longer has the significant added front-office provisioning and service operation. As was the case from the customer standpoint, a gamma service is less environmentally taxing from the standpoint of the service provider.

It is perhaps worth pointing out that in some cases a service can be accomplished with either an alpha or a beta approach: Household recycling can occur either by pickup or dropoff, for example, and haircuts may be given by a barber in his facility or the customer's. Many non-environmental considerations enter into decisions on how to provide these services, but the environmental attributes of the different approaches might well be factors that enter into the considerations.

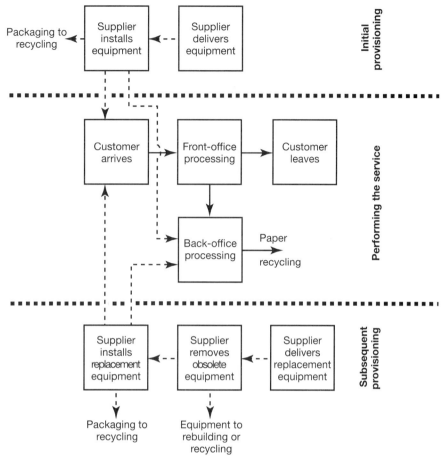

FIGURE 10.5 Continued.

10.8 THE SERVICES ECONOMY

On one level, the SLCA procedures described in this chapter may be viewed as convenient ways in which to approach assessing the resource use and environmental impacts of a sector of societal activity distinct from manufacturing per se. Such a viewpoint is legitimate and utilitarian, but may insufficiently far-sighted. Indeed, there are increasing demonstrations that the service industries may completely dominate the interactions among materials, products, and consumers. Consumers value products because the products provide services: food provides nutrition (or pleasure), vacuum cleaners provide cleaning services. This suggests that the difference between products and services is rather narrow with regard to the utility that they provide. Indeed, some researchers have proposed the perspective of a *utilization economy*, in which one purchases not a product but a function. Operative ways by which these activities can occur are already with us in the form of leases for photocopy machines and refundable

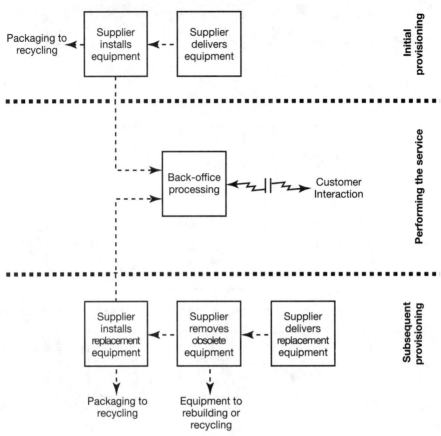

FIGURE 10.5 Continued.

deposits on beverage containers. A concomitant benefit for society of the utilization economy is that the producer, not the consumer, would have the primary responsibility for disassembly and recycling of products and their materials, and materials flow loops would begin to close.

As the percentage of societal actions embodied by services continues to increase, the environmental responsibility of service providers will become increasingly important. The environmental impact of service operations has generally been overlooked, while the more visible manufacturing operations have begun to be addressed. This situation is particularly unfortunate in that service industries are numerous and pervasive and their total impacts are very large. As with other sectors of society, it is the duty of the service provider to examine his operations and determine how to make them environmentally responsible. The assessment tools given in this chapter provide a place for service industries to begin in accomplishing their part of the vision of an environmentally responsible technological society.

FURTHER READING

Chouinard, Y., and M. Brown, Going organic: Converting Patagonia's cotton product line. *Journal of Industrial Ecology, 1* (1997) 117–129.

Cramer, J.M., and A.L.N. Stevels, A model for the take-back of discarded consumer electronics products. *Clean Electronics Products and Technology*, Conf. Pub. 415 (Institution of Electrical Engineers, London, 1995) pp. 129–141.

DeKeyser, D.M., and D.A. Eijadi, Development of the Andersen lighthouse for the WalMart environmental demonstration store. In *Proc. 2nd International Green Building Conference*, Special Pub. 888 (Natl. Inst. of Standards and Technology, Gaithersburg, MD, 1995) pp. 143–151.

Hopkins, L., D.T. Allen, and M. Brown, Quantifying and reducing environmental impacts resulting from transportation of a manufactured garment. *Pollution Prevention Review, 4* (1994) 491–500.

Lober, D.J., and M.D. Eisen, The greening of retailing. *Journal of Forestry, 93* (1995) 38–41.

Stahel, W., The utilization-focused service economy: Resource efficiency and product-life extension, in *The Greening of Industrial Ecosystems*, edited by B.R. Allenby and D.J. Richards (Washington, D.C.: National Academy Press, 1994) pp. 178–190.

EXERCISES

10.1. Choose a service facility with which you are familiar, perhaps because you have been employed in helping to provide the service or because you use the service frequently. Classify the service as alpha, beta, or gamma, and describe the product provided by the service facility and the process used to provide it.

10.2. Using the guidelines and protocols of Appendix D, evaluate where possible each matrix element (Table 10.4) for the service facility of Exercise 10.1. For those matrix elements where you could not perform an evaluation, what additional information was needed? How would you go about getting it?

10.3. Draw the target plot for the matrix completed in Exercise 10.2, omitting points where you were unable to assign a matrix element value. What life stages and what stressors are of most concern?

10.4. What recommendations would you make for improving the environmental attributes of the service facility chosen in Exercise 10.1? How would you prioritize the recommendations and what are the bases for the prioritization?

CHAPTER 11
Integrated Business Unit, Corporate, and Intercorporate Assessment

11.1 INTRODUCTION

The previous chapters have presented techniques for assessing the degree of environmental responsibility of individual products, processes, facilities, and services. A higher level of assessment is to integrate several of these approaches in order to evaluate all products in a product line, for example, or all facilities within a business unit. The reason for performing integrated assessments is that they will often reveal problems common to more than one activity, such as a lack of attention to packaging no matter what the product or a consistent use of pesticides and herbicides on all facilities in an area. This chapter explores techniques for performing such integrated assessments.

11.2 INTEGRATED PRODUCT ASSESSMENTS

Imagine a product line composed of p products, each of which has been assessed by the matrix techniques of Chapter 7. For each product, there are 25 matrix element evaluations. An integrated assessment of the product line can be performed directly by merely averaging the individual matrix element values:

$$P_{i,r} \text{ (composite)} = \frac{\sum_p P_{i,r,p}}{p} \qquad (11.1)$$

An example of an integrated study of this type is an assessment performed by the author on nearly two dozen AT&T telecommunications products. The products included those as small and inexpensive as electronic connectors and as large and costly as electronic switching machines, as well as representations of the offerings of several different business units. When the composite calculations were performed, they yielded the ratings shown in Table 11.1. (The comments refer to the principal factors responsible for reducing the matrix element values from the maximum score of 4. No attempt has been made to apply weighting factors based on product sales volume, cost, or other "external" factors.)

TABLE 11.1 An Environmental Assessment of AT&T Corporate Products

Element Designation	Element Value and Explanation
Premanufacture	
Matls. choice (1,1)	2.3 (Recycled materials seldom used.)
Energy use (1,2)	2.2 (Substantial energy use in intercontinental transport.)
Solid res. (1,3)	2.0 (Little attempt to minimize packaging volume or diversity.)
Liquid res. (1,4)	3.4 (Only minimal residue produced.)
Gas res. (1,5)	2.5 (Substantial emissions from intercontinental transport.)
Product Manufacture	
Matls. choice (2,1)	3.1 (Lead solder used, but waste lead tightly controlled.)
Energy use (2,2)	3.1 (Wave-solder machines use significant amounts of energy.)
Solid res. (2,3)	2.8 (Some solder paste residues produced, but effectively handled.)
Liquid res. (2,4)	3.3 (Low-solids flux residues must be chemically neutralized.)
Gas res. (2,5)	3.3 (Only minimal residue produced.)
Product Delivery	
Matls. choice (3,1)	2.6 (Two materials generally used, one is preferred.)
Energy use (3,2)	3.5 (Only minimal energy use.)
Solid res. (3,3)	1.7 (No recycling instructions; some packaging not recyclable.)
Liquid res. (3,4)	4.0 (No residue produced.)
Gas res. (3,5)	3.9 (Only minimal residue produced.)
Customer Use	
Matls. choice (4,1)	3.7 (No significant concerns.)
Energy use(4,2)	3.3 (Products generally have modest energy use.)
Solid res.(4,3)	3.9 (Only minimal residue produced.)
Liquid res(4,4)	4.0 (No residue produced.)
Gas res.(4,5)	3.8 (Only minimal residue produced.)
Refurbishment, Recycling, Disposal	
Matls. choice (5,1)	2.1 (Lead present on printed circuit boards.)
Energy use(5,2)	2.1 (Significant disassembly costs for labor and energy.)
Solid res.(5,3)	1.7 (Designs offer little opportunity to reuse materials.)
Liquid res(5,4)	3.7 (Only minimal residue produced.)
Gas res.(5,5)	3.6 (Only minimal residue produced.)

If these scores are summed over all life stages and all environmental stressors, the result is an overall rating for the telecommunications products of 75.6, certainly commendable, but also leaving plenty of room for improvement.

The corporate target plot generated from the composite assessment is shown in Figure 11.1. It indicates that the manufacturing processes are generally exemplary from a DFE standpoint, although room still exists for minimizing solid residues and water use. Also on the positive side is the minimal level of environmental impacts during product use, although energy consumption might be further examined. In contrast to these favorable attributes, several corporate-wide DFE product characteristics tend to be less than completely satisfactory. These include impacts from the use of virgin materials, the volume and diversity of supplier packaging entering the manufacturing facilities, the environmental impacts (energy use and gaseous emissions) of intercontinental shipping of components and products, the volume and diversity of packaging used to ship the products, the use of lead-containing solder, and minimal efforts to include design for disassembly characteristics in the products.

172 Chapter 11 Integrated Business Unit, Corporate, and Intercorporate Assessment

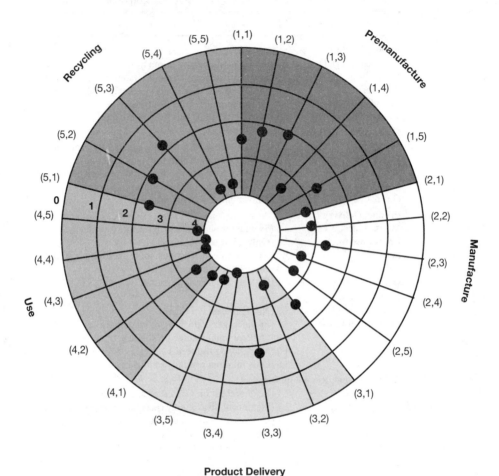

FIGURE 11.1 A composite target plot for telecommunications products.

The results of the composite DFE assessment suggest a number of recommendations that will make all corporate telecommunications products increasingly preferable from an environmental standpoint. Among the more important of these were the following recommendations for physical design organizations:

- Improve the modularity of the products.
- Consider redesign of subassemblies and components so that they can be joined with quick-release fasteners and the like rather than screws.
- Minimize plastics diversity, especially the use of pigmented or other difficult to recycle materials.
- Mark all plastic parts using International Standards Organization guidelines.

Important manufacturing organizations:

Important manufacturing organization recommendations:

- Rewrite specifications for components used in the products to specify the use of increased recycled material in their manufacture.
- Minimize the diversity of materials in outgoing product packaging, and develop labels to affix to the packaging to indicate appropriate recycling procedures to the customer.

Important supply line management organization recommendations:

- Work with suppliers to minimize the diversity of packaging material entering the manufacturing facilities so that recycling of solid residues may be optimized.
- Rewrite specifications for incoming packaging material to specify the use of some recycled material in its manufacture.
- Because transport (especially intercontinental transport) of components and products carries environmental impacts related to energy use and gaseous emissions, the most energy efficient routes and most environmentally benign transport technologies should be used.

Assessments of several processes within a facility or of several facilities can be performed in the same way. As before, each process or each facility is evaluated individually. The integrated matrix element values are then given for multiple processes:

$$Q_{j,r} \text{ (composite)} = \frac{\sum_q Q_{j,r,q}}{q} \quad (11.2)$$

where j is the life stage index, r the environmental stressor index, and q the process index. For multiple facilities, the relationship is

$$F_{k,r} \text{ (composite)} = \frac{\sum_f F_{k,r,f}}{f} \quad (11.3)$$

where k is the life stage index, r is the environmental stressor index, and f is the facility index.

11.3 INTEGRATED OPERATIONS ASSESSMENTS

In addition to integrated assessments of a family of products, processes, or facilities, a manager may want to make an assessment of all the operations for which she or he is responsible. In such a case, the interaction among products, processes, and facilities must be considered, as shown on the overlap plot of Figure 11.2. In principal, Figure 11.2 need not represent a single product, a single process, and a single facility, but any number of each. Most of the life stages are distinct, but a common stage (marked C on the figure) occurs where product manufacture, principal process and complementary process operation, and facility product and process realization overlap.

To demonstrate how a combined integrated assessment is performed, picture an operation encompassing p products made by q processes (similar or identical

174 Chapter 11 Integrated Business Unit, Corporate, and Intercorporate Assessment

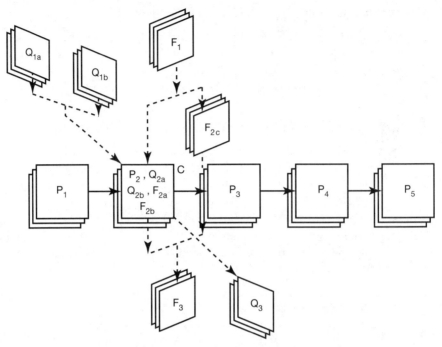

FIGURE 11.2 Life stage diagram for an integrated assessment of products (P), processes (Q), and facilities (F). The common intersection area is designated C, and the subscripts refer to the life stages defined in Figs. 2.1, 8.2, and 9.1. The spectrum of environmental stressors applies to each life stage, so there is an additional dimension not seen in the diagram. As indicated, multiple products, processes, and facilities can be included.

processes for each product) at f facilities around the world. Assume that for each product, process, and facility a matrix assessment has been performed. One begins the integrated assessment by taking simple arithmetic averages of the results of each life stage except for the common stage C:

$$P_{i,r} \text{ (composite)} = \frac{\sum_p P_{i,r,p}}{p} \text{ for i} = 1,3,4,5 \tag{11.4}$$

$$Q_{j,r} \text{ (composite)} = \frac{\sum_q Q_{j,r,q}}{q} \text{ for j} = 1a,1b,3 \tag{11.5}$$

$$F_{k,r} \text{ (composite)} = \frac{\sum_f F_{k,r,f}}{f} \text{ for k} = 1,2c,3 \tag{11.6}$$

Common life stage C is evaluated by

$$C_r = \frac{\sum_p P_{2,r,p} + \sum_q Q_{2a,r,n} + \sum_q Q_{2b,r,n} + \sum_f F_{2a,r,f} + \sum_f F_{2b,r,f}}{p + 2q + 2f} \quad (11.7)$$

The maximum assessment score for an individual product, process, or facility is 100. Because of normalization in equations 11.1 to 11.6, the same is true of the composite scores. Integrated operations assessments, which consider products, processes, and facilities simultaneously, thus have a maximum score of 300. The actual overall operations assessment score is given by

$$M_{OP} = \sum_{i=1,3-5} \sum_{r=1,5} P_{i,r}(\text{composite}) + \sum_{j=1a,1b,3} \sum_{r=1,5} Q_{j,r}(\text{composite})$$
$$+ \sum_{k=1,2c,3} \sum_{r=1,5} F_{k,r}(\text{composite}) + 5\sum_{r=1,5} C_r \quad (11.8)$$

where C is multiplied by 5 to reflect its representation of 5 of the 15 life stages in the integrated assessment. As has been noted earlier in this book, the most important result of performing such an assessment is almost surely not the score derived, but the recommendations for improved performance that result.

The approach outlined above applies to the situation in which an integrated approach to the optimization of raw materials flow streams and residue flow streams has been undertaken, that is, where activities within a single manufacturing building, manufacturing complex, or corporate manufacturing organization have been treated as independent entities. In some cases, however, especially in the chemical industry, by-product flows from some processes are designed to be utilized as input flows to others; these flows may be materials, heat, combustible waste products, or any other useful resource. In such a case, a residue stream from process A that would receive a low matrix element rating if consigned for disposal would likely receive a higher rating if utilized by process B. Similarly, the process B rating, which would have been low had virgin raw materials been utilized, would likely have a higher rating when advantage is taken of the availability of the process A residue stream. To perform an integrated operations assessment where residue streams are shared, therefore, the matrix assessments for those operations involved in stream sharing must be redone from the shared resource perspective.

11.4 INTEGRATED INTERCORPORATE ASSESSMENTS

It is increasingly common for governments or alliances of corporations to establish "industrial parks," which are areas set aside for industrial water and sewer networks and road and rail systems. In such a setting, one can also envision the creation of "eco-parks," in which industrial activities are designed by the participants so that the residue streams of some facilities become resource streams for others, much as might happen in the integrated facility of a single corporation.

The textbook example of an ecopark, indeed perhaps the only true ecopark example to date, is that of Kalundborg, Denmark. Four main participating corporations are involved: the Asnaes Power Company, a Novo Nordisk pharmaceutical plant, a Gyproc facility for producing wallboard, and a Statoil petroleum refinery. The secondary participants include a sulfuric acid plant, a road surface contractor, a flower grower, a fish farm, a number of field crop farmers, and the local community heating utility. Individually, the resource and residue streams of each of the principal participants are shown in Table 11.2 and diagrammed in Figure 11.3. The ecopark's participants do not function individually, however, but as a symbiotic whole, as shown in Figure 11.4. Steam, gas, cooling water, and gypsum are exchanged among the participants, and some heat also is used for fish farming and home and greenhouse heating. The residual products—sulfur, fly ash, and sludge—not usable in the immediate vicinity are sold for use elsewhere. None of the arrangements was required by law; rather, all were negotiated independently for reasons of better materials prices or avoidance of materials disposal costs.

TABLE 11.2 Materials and Energy Flows in the Kalundborg Industrial Ecosystem.

Resources	Products	Residues
Asnaes		
Coal	Electrical power	Gypsum[2]
		Steam[2]
Gas[1]		Heat[2]
Oil		Fly ash[2]
		Combustion gases
Water[1]		
Statoil		
Crude oil	Petrol	Gas[2], sulfur[2]
	Refinery products	Oily waste
Water		Wastewater[2]
Steam[1]		
Electricity[1]		Heat
Gyproc		
Gypsum[1]	Wallboard	
Electricity[1]		Heat
Water		Wastewater
Gas[1]		
Cardboard		
Novo Nordisk		
Water	Insulin	Sludge[2]
Electricity[1]	Ind. enzymes	
Steam[1]		

[1] These resources are obtained wholly or in part from other members of the Kalundborg industrial ecosystem.
[2] These residues are regarded as by-products rather than wastes; they are not discarded, but are transferred to others within or outside the Kalundborg industrial ecosystem.

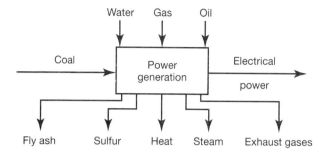
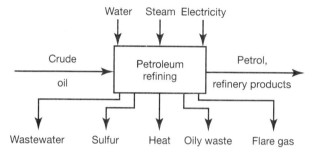
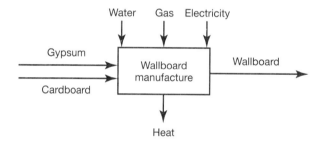
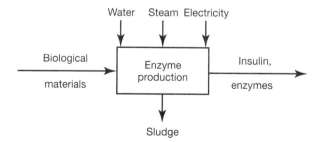

FIGURE 11.3 The principal industrial operations at Kalundborg, Denmark, treated as individual facilities.

The environmental assessment of such an ecopark is performed the same as an assessment for an intercorporate facility that shares resources among different processes. Thus, each corporate entity in an ecopark is individually evaluated, with the resource sharing taken into account. The ecopark assessment is then performed by utilizing the summation techniques described in this chapter. For brevity, individual corporate assessments for Kalundborg will not be reproduced here. Rather, target plots

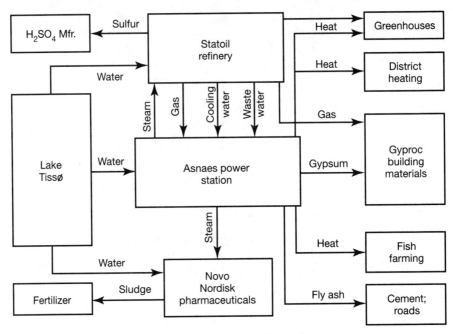

FIGURE 11.4 The industrial ecosystem at Kalundborg, Denmark.

representing the integrated intercorporate assessment are given in Figure 11.5 for two situations: as if the Kalundborg facilities were independent (and could be located anywhere) and as they are in an industrial ecosystem.

The target plots demonstrate that the assessments of facility development and closure are unaffected by whether or not the facilities are part of an industrial ecosystem. Also unaffected is the assessment of products, because those are influenced by customers and internal decisions, not by neighbors in other businesses. Facility operations may be slightly affected because members of an ecosystem may cooperate on certain common activities such as the recycling of solid and liquid wastes. The dramatic difference, however, occurs in the processes stage. Here what formerly were liabilities —loss of heat, emission of acid gases, ocean dumping of sludge, to name a few—have become benefits, because those lost energy and materials have become useful by-products. One's initial reaction to this analytical result might be dismay that only one of the five life stages is greatly affected by the industrial ecosystem approach, but it is perhaps the most important life stage, because environmentally unsound processes go on having detrimental impacts day after day and year after year if not addressed.

11.5 AN INTEGRATED ASSESSMENT PERSPECTIVE

The integrated assessments discussed in this chapter are difficult to defend in terms of rigor; they deal with material flows of different magnitudes, notable impacts on differ-

Section 11.5 An Integrated Assessment Perspective 179

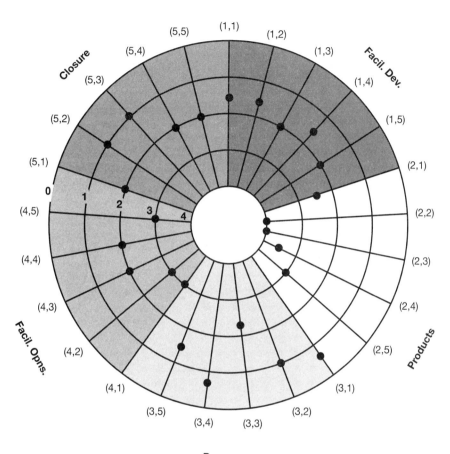

FIGURE 11.5 (a) The composite target plot for the Kalundborg ecopark computed as if the facilities were totally distinct from one another; (b) the intercorporate target plot for the Kalundborg complex computed by taking into account its industrial ecosystem characteristics.

ent spatial scales, and consideration of short, intermediate, and long time spans. The concept of the functional unit is subsumed in the individual assessments, and merging those assessments blurs the utility of the approach. The result is perhaps more accurately an extended environmental audit than an SLCA.

Notwithstanding these difficulties, however, an integrated assessment can yield substantial environmental benefits. Without necessarily being able to quantify the expected gains, corporations and groups of corporations can nonetheless direct overall efforts at residue reduction, packaging improvements, or other general activities. The results will inevitably be beneficial to the environment and to the corporate or intercorporate entities involved.

180 Chapter 11 Integrated Business Unit, Corporate, and Intercorporate Assessment

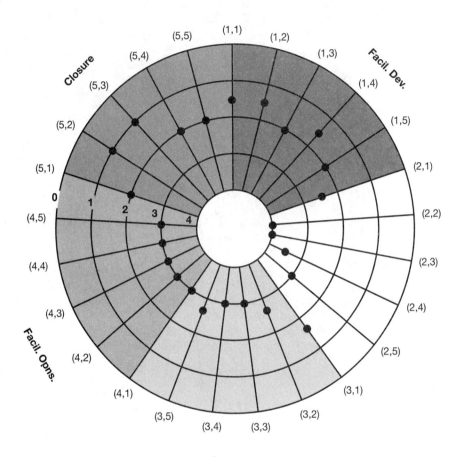

FIGURE 11.5 Continued.

FURTHER READING

Gertler, N., *Industrial Ecosystems: Developing Sustainable Industrial Structures* (Unpublished Master's Thesis, Massachusetts Institute of Technology, Cambridge, MA, , 1995).

Graedel, T.E., P. Comrie, and J.C. Sekutowski, Green product design, *AT&T Technical Journal*, 74 (6) (1995) 17–25.

EXERCISES

11.1. Assume that you are the manager of a chain of five automobile repair shops and you want to assess the environmental responsibility of your operation. How might Figure 11.2 be changed to take into account the fact that you are providing services, not manufacturing products?

11.2. The multinational corporation for which you work has manufacturing facilities in Sweden, Italy, Malaysia, Australia, and Mexico. You are asked to perform an integrated operations assessment for the corporation. What issues need to be considered that would not arise if all manufacturing was done in Sweden?

11.3. You have completed an intracorporate environmental assessment for your corporation, and the results are favorable. How should such an assessment be used with employees? With the Board of Directors? With shareholders? With outside interest groups? Defend your answer and describe the ways in which the information sharing might differ with different groups.

11.4. Refer to the industrial ecosystem in Figure 11.4. If the wallboard manufacturing facility and the cement plant were located 50 km away from the other facilities but still wanted to share input-output streams, how would the ecosystem be likely to be affected? What factors would determine whether the arrangement was still attractive?

CHAPTER 12
Assessment of Societal Infrastructure

12.1 INFRASTRUCTURE SYSTEMS

An infrastructure system is a physical service network provided by society to its citizens, generally through a combination of several public and/or private sector entities. An example is the system in most communities that supplies drinking water to all residents, commercial establishments, and industrial facilities and removes and treats their wastewater. In many parts of the world, especially in countries whose industrial systems are highly developed, infrastructure systems, or at least provisions to enable such systems to organize and operate, are assumed to be a part of the responsibilities of government.

Infrastructure systems as such have seldom or never been the subject of life-cycle assessments, although the environmental impacts of individual components of those systems have been. Entire infrastructure systems turn out to be particularly appropriate for environmental impact analyses from a societal standpoint, because infrastructure is enabled by society and the resulting environmental impacts are imposed upon that same society, at least if complications of time and space are overlooked.

Infrastructure systems can be defined in different ways, but at least some systems are so common and obvious as to have close to universal recognition. They include:

- The electrical power system
- The water and sewer system
- The transportation system
- The solid waste management system
- The telecommunications system
- The national defense system
- The public safety system
- The medical care system

The thoughtful reader will notice that some of these systems are quite distinct (e.g., water and sewer) and others (e.g., national defense and transportation) are intertwined. This causes no complications as long as the boundaries of the infrastructure study are carefully defined. In some countries, several of these systems are operated by private entities, but governments generally exercise regulatory oversight.

Subdivisions of the list of infrastructure systems are not only possible, they are frequently performed. For example, the transportation system may be treated as a unit, or its subdivisions—the automotive infrastructure system, the rail infrastructure system, and the air transport infrastructure system—may be treated separately. Infrastructure assessments may also be done on different spatial scales; one could, for example, look at the road transportation system of a town, a region, a country, or a continent.

Infrastructure system life-cycle assessment moves LCA/SLCA from the corporate realm to the societal realm. It is, in some sense, a generalization and expansion of the environmental impact assessments done in many countries when a large public project is undertaken. Hence, this chapter is primarily directed to the public sector, and to those corporations that interact with the public sector.

12.2 MATRIX APPROACHES FOR INFRASTRUCTURE SYSTEMS

A suitable assessment system for environmentally responsible infrastructure systems (ERISs) should treat the following environmental stressors: impacts on biodiversity, choice of materials, energy use, and generation of solid, liquid, and gaseous residues of various types. As with the previous approaches in this book, the assessment system can be structured as a 5×5 matrix, one dimension of which is life-cycle stages and activities and the other of which is environmental stressor. The five life stages for infrastructure are illustrated in Figure 12.1 and are defined in the following sections.

Stage 1: Site Development

A significant factor in the degree of environmental responsibility of a societal infrastructure is the site or sites selected and the way in which development is planned and carried out. Minimization of ecosystem disturbance is among the desirable activities associated with this life stage.

Stage 2a: Materials and Product Delivery

A major potential environmental impact of infrastructure development is the methods used to deliver materials and products to the construction site. Energy use should be minimized, if possible, by limiting the transport of heavy materials. Solid waste generation should be reduced by detailed planning and purchasing to minimize construction debris.

Stage 2b: Infrastructure Manufacture

This stage treats all aspects of the construction of the infrastructure itself, including the choice of materials, their delivery to the site, techniques and equipment used in construction, and site cleanup.

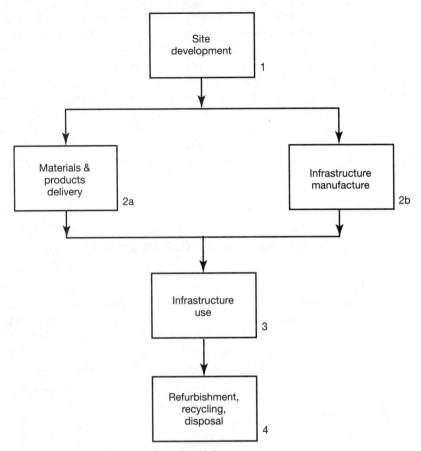

FIGURE 12.1 The life stages of a societal infrastructure system.

Stage 3: Infrastructure Use

Some infrastructure systems have minimal in-use impacts, while the impact of others is substantial. The usual items treated in this life stage are maintenance and repair of infrastructure components and the use of energy as the system operates.

Stage 4: Refurbishment, Recycling, Disposal

Just as environmentally responsible products and facilities are increasingly being designed for product life extension, so should societal infrastructures. To the degree possible, infrastructures should be designed to be easily refurbished and upgraded. If they must be retired, the design should permit recovery of materials, fixtures, and other components for reuse or recycling.

The assessment matrix for societal infrastructure is shown in Table 12.1. To preserve consistency among the assessment techniques presented in this book, each element of the matrix is assigned an integer rating from 0 (highest impact, a very negative evaluation) to 4 (lowest impact, an exemplary evaluation). The checklists and guidelines of Appendix E are used as aids to the assessor.

TABLE 12.1   The Environmentally Responsible Infrastructure System Matrix

	Environmental Stressor				
Facility Activity	Ecological Impacts	Energy Use	Solid Residues	Liquid Residues	Gaseous Residues
Site development	1,1	1,2	1,3	1,4	1,5
Materials and product delivery	2,1	2,2	2,3	2,4	2,5
Infrastructure manufacture	3,1	3,2	3,3	3,4	3,5
Infrastructure use	4,1	4,2	4,3	4,4	4,5
Refurbishment, recycling, disposal	5,1	5,2	5,3	5,4	5,5

(The numbers are the matrix element indices i, j.)

After an evaluation has been made for each matrix element, the overall Environmentally Responsible Infrastructure System Rating (R_{ERIS}) is computed as the sum of the matrix element values:

$$R_{ERIS} = \sum_i \sum_j M_{i,j} \qquad (12.1)$$

Because there are 25 matrix elements, a maximum infrastructure system rating is 100.

12.3   ASSESSING THE AUTOMOTIVE INFRASTRUCTURE OF YESTERDAY AND TODAY

As with the automobile, the automotive infrastructure has environmental impacts during all its life stages, and these impacts can be evaluated by the streamlined assessment tools described above. Like the automobile assessment, it is instructive to perform environmentally responsible systems assessments on examples of generic automotive infrastructure of the 1950s and 1990s: a unit length of high-speed highway. Here the functional unit is the manufacture and maintenance of a portion of highway infrastructure. The magnitude of the unit length is not critical, because only qualitative aspects will be considered, but one can think in terms of 5–10 km as a unit length that would encompass all relevant aspects of the roadway and its related components. Some of the characteristics of the generic infrastructure systems are given in Table 12.2.

In overview, the 1950s infrastructure was a multi-lane highway representing a transition of an older two-lane roadway into a higher speed, unlimited-access roadway. The road widening needed to accomplish the transition was unconstrained by environmental considerations, as was the subsequent asphalt paving and roadway maintenance. Small amounts of steel were used for such ancillary purposes as signs and occasional guard rails. Patching and other surface repair was performed as needed.

The 1990s infrastructure, in contrast, is a limited-access highway of the type represented by the German Autobahn, the Italian Autostrada, or the U.S. Interstate Highway. The surface is made of reinforced concrete, as are the bridges and overpasses,

TABLE 12.2 Characteristics of Automotive Infrastructure of the 1950s and 1990s

Characteristic	ca. 1950s	ca. 1990s
Roadway Characteristics		
Type	Unlimited access	Limited access
Speed limit	90 km/hr	110 km/hr
Surface	Asphalt	Reinforced concrete
Route	Historic road	Farmland, wetland
Site Development		
Location	Unconstrained	Moderate control
Solid residues	No control	Minor control
Energy use	Very high	Very high
Infrastructure Manufacture		
Stone	Heavy use	Heavy use
Minerals	Light use	Heavy use
Petroleum	Heavy use	Moderate use
Energy use	Very high	Very high
Materials and Product Delivery		
Component packaging	Minor amount	Large amount
Packaging recycling	No	Minor amount
Infrastructure Use		
Energy use, patching	High	High
Solid residues, patching	No control	Minor control
Painting	Pb, high VOC	Pb, moderate VOC
Road sand, salt	Minor amount	Large amount
Infrastructure Recycling		
Asphalt	None	Modest amount
Concrete	None	Minor amount
Steel	Small amount	Substantial amount

and extensive signage and overhead lighting are used at roadway exits. To make the roads safer, concrete barriers are extensive and entrance and exit ramps are long, in a sense trading safety for ecosystem disturbance. Drainage control is fairly extensive, both during construction and during use. Sand and salt are spread extensively on the roadway in the winter months to improve traction and prohibit ice formation. Every several years the roadway surface is removed and new roadway is laid down.

A significant factor in the degree of environmental responsibility of an infrastructure system is the sites selected and the way in which those sites are developed. A large fraction of the transportation infrastructure is constrained to be located in or near urban areas. For infrastructure of any kind built on land previously undeveloped as industrial or commercial sites, ecological impacts on regional biodiversity can be anticipated, as well as added air emissions from new transportation and utility infrastructures. These effects can be minimized with attention to working with existing infrastructures and developing the site with the maximum area left in natural form. Nonetheless, given the requirement for infrastructure to follow need, there is a strong societal component in how that infrastructure is developed. The ratings assigned to the first life stage of generic infrastructure from each epoch are given in Table 12.3, and the

TABLE 12.3   Site Development

Element Designation	Element Value and Explanation
1950s	
Biodiversity/materials (1,1)	0 (No concern for site effects.)
Energy use (1,2)	1 (Very high energy consumption.)
Solid residue (1,3)	0 (No solid residue control.)
Liq. residue (1,4)	1 (Minimal liquid residue control.)
Gas residue (1,5)	1 (Minimal gaseous residue control.)
1990s	
Biodiversity/materials (1,1)	1 (Large land areas altered.)
Energy use (1,2)	1 (Very high energy consumption.)
Solid residue (1,3)	1 (Limited solid residue control.)
Liq. residue (1,4)	2 (Minimal liquid residue control.)
Gas residue (1,5)	2 (Minimal gaseous residue control.)

TABLE 12.4   Materials and Product Delivery

Element Designation	Element Value and Explanation
1950s	
Biodiversity/materials (2,1)	3 (Benign, common materials used.)
Energy use (2,2)	2 (Manufacture and shipping is energy-intensive.)
Solid residue (2,3)	2 (Small amounts of packaging for signage.)
Liq. residue (2,4)	3 (Small amounts of liquids are generated by packaging and shipping.)
Gas residue (2,5)	2 (Moderate fluxes of gases produced in bitumen manufacture.)
1990s	
Biodiversity/materials (2,1)	3 (Benign, common materials used.)
Energy use (2,2)	1 (Manufacture and shipping is very energy-intensive.)
Solid residue (2,3)	2 (Moderate amts. of packaging for signage, lighting, etc.)
Liq. residue (2,4)	3 (Small amounts of liquids are generated.)
Gas residue (2,5)	1 (High fluxes of greenhouse gases produced by cement manufacture.)

two numbers in parentheses refer to the matrix element indices shown in Table 12.1. The higher (that is, more favorable) ratings for the 1990 infrastructure are mainly due to improvements in the environmental aspects of site development.

The environmental stressors at life stage 2a, materials and product delivery, have two components (Table 12.4). Where infrastructure is manufactured on-site, as with asphalt or concrete, the concerns encompass extraction, processing, and transport of the materials delivered to the site and used in the manufacture of pavements or structures. Where infrastructure materials or components are manufactured elsewhere and consumed or installed on site, as with cement, bridges, girders, light stanchions, or signage, the concerns encompass manufacture of the material, packaging, transportation of the product to the customer, and product installation.

In the case of the 1950s roadway, the refinery manufacture of bitumen was energy-intensive and greenhouse gas emissions were high. Lights, signage, and the like were not used to a great extent, and were not overpackaged, so site residues from shipping and installation were modest.

The 1990s roadway of concrete required large volumes of cement, the manufacture of which is very energy intensive and which emits very large quantities of greenhouse gases. Infrastructure components manufactured off-site are abundant, and their packaging, shipping, and installation create significant impacts.

Life stage 2b is infrastructure manufacture (Table 12.5). Many of the basic automotive infrastructure manufacturing processes have changed little over the years, although modest efforts have been made to improve their environmental responsibility. The big difference is the progression from unlimited-access asphalt surfaces to limited-access highways constructed largely of reinforced concrete, and with overpasses, bridges, and entrance and exit ramps of the same material. Such a roadway requires more materials throughout; overall energy use is higher than for an asphalt surface, but the rates of gaseous emissions during construction are lower.

Life stage 3 comprises impacts from consumables or maintenance materials that are expended while keeping the automotive infrastructure well-maintained and functioning properly (Table 12.6). The two infrastructures differ dramatically. The 1950s asphalt surface required modest amounts of virgin materials to renew it. (Unlike today, asphalt was not then recycled.) Small amounts of sand and salt were used to maintain winter traction. The exhausts on the maintenance vehicles were uncontrolled.

The 1990s roadway requires frequent repair, largely due to use by heavy trucks, and is more difficult to maintain once it is degraded. The broken pavement or rusted reinforcing steel cannot be horizontally recycled, so new material is required. Road sand and salt are used extensively. Some emission control is in effect for maintenance vehicles.

The fourth life stage assessment includes impacts during infrastructure refurbishment or as a consequence of the eventual discarding of components or entire systems deemed obsolete and too costly to recycle (Table 12.7). Neither the 1950s roadway nor that of the 1990s is specifically designed with end of life in mind, but the asphalt surface is horizontally recyclable to other roadways or to itself; concrete is not. The reinforcing steel in the 1990s roadway is difficult to separate from the concrete matrix, and

TABLE 12.5 Infrastructure Manufacture

Element Designation	Element Value and Explanation
1950s	
Biodiversity/materials (3,1)	3 (Most materials are relatively benign.)
Energy use (3,2)	2 (High energy consumption.)
Solid residue (3,3)	2 (Substantial solid residues from materials processing.)
Liq. residue (3,4)	2 (Substantial liquid residues from materials processing.)
Gas residue (3,5)	2 (Moderate gaseous residues from asphalt manufacture.)
1990s	
Biodiversity/materials (3,1)	2 (Most materials relatively benign, but used prolifically.)
Energy use (3,2)	1 (Very high energy consumption.)
Solid residue (3,3)	2 (Substantial solid residues from materials processing.)
Liq. residue (3,4)	2 (Substantial liquid residues from materials processing.)
Gas residue (3,5)	3 (Modest gaseous residues from construction vehicles.)

Section 12.3   Assessing the Automotive Infrastructure of Yesterday and Today   189

TABLE 12.6   Infrastructure Use

Element Designation	Element Value and Explanation
1950s	
Biodiversity/materials (4,1)	2 (Minor road salting, little matls. reuse.)
Energy use (4,2)	2 (Fossil fuel energy use in maintenance is large.)
Solid residue (4,3)	3 (Sand, degraded pavement are residues.)
Liq. residue (4,4)	3 (Modest fluid emissions.)
Gas residue (4,5)	1 (High maintenance vehicle exhaust emissions.)
1990s	
Biodiversity/materials (4,1)	0 (Extensive road salting, no matls. reuse.)
Energy use (4,2)	1 (Fossil fuel energy use in maintenance is large.)
Solid residue (4,3)	1 (Sand, salt, degraded pavement, steel.)
Liq. residue (4,4)	3 (Modest fluid emissions.)
Gas residue (4,5)	2 (Substantial maintenance vehicle exhaust emissions.)

TABLE 12.7   Refurbishment/Recycling/Disposal

Element Designation	Element Value and Explanation
1950s	
Biodiversity/materials (5,1)	3 (Most materials used are recyclable.)
Energy use (5,2)	2 (Substantial energy use required to recycle materials.)
Solid residue (5,3)	3 (Some materials and components discarded.)
Liq. residue (5,4)	4 (Liquid residues from disposal are minimal.)
Gas residue (5,5)	3 (Gaseous residues from recycling or disposal are moderate.)
1990s	
Biodiversity/materials (5,1)	3 (Materials used are difficult to recycle.)
Energy use (5,2)	2 (Substantial energy use required to recycle materials.)
Solid residue (5,3)	2 (Many materials and components discarded.)
Liq. residue (5,4)	4 (Liquid residues from final life stage are minimal.)
Gas residue (5,5)	3 (Gaseous residues from final life stage are moderate.)

tends to be corroded by road salt over time; it thus has little value. Signage, lighting stanchions, and other steel components can generally be recycled.

The completed matrices for the generic 1950s and 1990s automotive infrastructure are illustrated in Table 12.8. Examine first the values for the 1950s infrastructure so far as life stages are concerned. The column at the far right of the table shows moderate environmental stewardship during most life stages. The ratings during site selection and preparation are very poor; no life stages demonstrate truly responsible performance. The overall rating of 53 is far below what might be desired. In the case of the 1990s infrastructure, product delivery, roadway manufacture, and end-of-life ratings are only passable, and the life stages of site selection and preparation and infrastructure use show poor environmental performance. The overall rating is 47, even lower than that of the earlier epoch. The principal reasons for these low ratings are the high degree of ecosystem disturbance during site preparation, severe environmental impacts during infrastructure use, and almost complete failure to consider optimizing

TABLE 12.8 Environmentally Responsible Assessments for the Generic 1950s and 1990s Automotive Infrastructure *

Life Stage	Environmental Stressor					
	Biodiversity/ Materials	Energy Use	Solid Residues	Liquid Residues	Gaseous Residues	Total
Site development	0	1	0	1	1	3/20
	1	1	1	2	2	7/20
Materials and product delivery	3	2	3	3	2	13/20
	3	1	2	3	1	10/20
Infrastructure manufacture	3	2	2	2	2	11/20
	2	1	2	2	3	10/20
Infrastructure use	2	2	3	3	1	11/20
	0	1	1	3	2	7/20
Refurbishment, recycling, disposal	3	2	3	4	3	15/20
	2	2	2	4	3	13/20
Total	11/20	9/20	11/20	13/20	9/20	53/100
	8/20	6/20	8/20	14/20	11/20	47/100

*Upper numbers refer to the 1950s infrastructure, lower numbers to the 1990s infrastructure.

the end of life of the roadways when developing the materials properties and infrastructure designs. The target plots of Figure 12.2 show opportunities for improvement at all life stages and over all environmental stressors.

12.4 DISCUSSION

The DFE situation with the road infrastructure is interesting and complex. Modern high-speed highways are undeniably successful in moving more vehicles than did the roads of a half-century ago, and they do so at significantly higher average speeds. However, from an environmental standpoint, the modern highway is not only not exemplary, but may be even worse than yesterday's poorer performing version. There is no doubt that the highway engineer faces a difficult balancing act in optimizing the sometimes conflicting demands of cost, roadway performance, and pressures for accomplishing improvements within an existing infrastructure rather than starting from scratch. Nonetheless, almost no effort has been given to the development of alternative roadway construction materials, recycling improvements, and transformations of in-use aspects. Much remains to be done to make the roadway infrastructure more environmentally responsible. This same scenario is repeated throughout society's infrastructures, and demonstrates the inability of societies and their designers to approach infrastructure from a systemic point of view.

FURTHER READING

Allen, D.T., Pollution Prevention: Engineering design on macro-, meso-, and microscales, *Advances in Chemical Engineering, 19* (1994) 251–323.

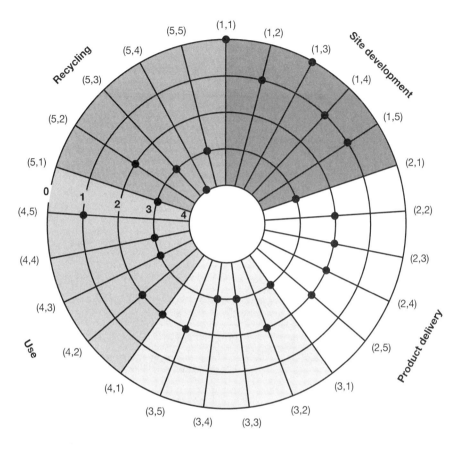

FIGURE 12.2 Comparative target plots for the display of the environmental impacts of the generic automotive infrastructure of the 1950s and of the 1990s.

Graedel, T.E., and B.R. Allenby, *Industrial Ecology and the Automobile* (Upper Saddle River, New Jersey: Prentice Hall, 1997).

EXERCISES

12.1. The medical care system is an infrastructure system with which most people are at least moderately familiar. Construct a table along the lines of Table 12.2, listing typical characteristics of the medical infrastructure of the 1950s and 1990s. Include all aspects of the system: hospitals, physician's offices, pharmacies, etc.

12.2. Repeat Exercise 12.1 for the U.S. national defense system.

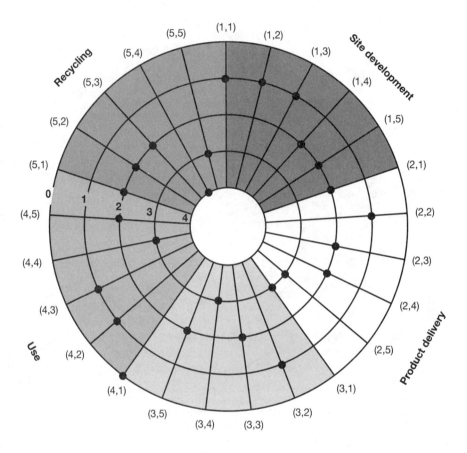

FIGURE 12.2 Continued.

12.3. Using the guidelines and protocols of Appendix E, evaluate, where possible, each matrix element (Table 12.1) for the infrastructure system of Exercise 12.2. For those matrix elements where you could not perform an evaluation, what additional information was needed? How would you go about getting it?

12.4. Draw the target plot for the matrix completed in Exercise 12.3, omitting points where you were unable to assign a matrix element value. What life stages and what stressors are of most concern?

12.5. What recommendations would you make for improving the environmental attributes of the infrastructure system of Exercise 12.2? How would you prioritize the recommendations and what are the bases for the prioritization?

CHAPTER 13

Upgraded Streamlining

13.1 THE UPGRADING CONCEPT

As assessors have grappled with the difficulties of detailed LCA and considered the merits and demerits of SLCA schemes, the recurring question—"What are we throwing away by using streamlined LCAs?"—is often followed by "Can we retain some of the simplicity and efficiency of SLCAs while using more of the extensive knowledge of environmental science?" The full answers to these questions remain to be supplied, but a few responses suggest themselves, responses that may ultimately prove to be particularly useful as SLCA methods move to software systems where additional complexity can be readily incorporated.

This chapter discusses techniques for augmenting and upgrading SLCAs. The first technique is a way in which to reflect positive benefits of projects that lower environmental impacts below what they were before the project was undertaken. The next is an explicit technique to incorporate localization into an SLCA. Finally, two techniques are presented to include valuation in an SLCA: one that applies weighting factors on the basis of a consensus of the assessment team, and one that relates weighting factors directly to the grand objectives of society.

13.2 BONUS SCORING

In the SLCA matrices that have been a central feature of this book, the standard approach to scoring has been to assign to each element of the matrix an integer rating from 0 (highest impact) to 4 (negligible impact) in an effort to reflect negative environmental impacts. Under unusual circumstances, however, a product, process, or facility may actually have a positive effect on the environment, and one might want to consider bonus scoring so that, say, a small positive effect would produce a rating of 5 and a large positive effect a rating of 6. An example of such circumstances is the recent proposal to plate automobile radiators with a thin film of platinum. Platinum is an efficient catalyst, and it decomposes ozone passed over it into molecular oxygen. Thus, a large fleet of vehicles with catalytic radiators operating in a smoggy urban environment could decrease ozone concentrations below what they would otherwise be.

Other examples come readily to mind once the concept of SLCA bonus scoring is appreciated. Facilities for the manufacture of electronic integrated circuits, for example, require highly purified water. Water is generally taken from municipal supply systems, purified by filtration and other techniques, used in the manufacturing process, and returned to municipal water infrastructures in purer condition that when it was initially acquired. The facility use of water in electronics manufacture thus deserves a bonus rating.

A more general look at the environmental stressors that have been discussed previously provides other ideas about what characteristics might constitute benefits:

- *Materials choice.* In lieu of using a readily obtainable virgin or recycled material, it may be possible to use a material for which no previous technological use was known and which would otherwise have been discarded.
- *Biodiversity.* Some activities, particularly in the landscape maintenance area, can result in the restoration of previously degraded land, thus providing more suitable biological environments than existed before the activity was begun.
- *Energy consumption.* Any activity that generates a net production of energy rather than a net consumption justifies bonus scoring. An example is a chemical process whose excess energy is used to generate power and provide that power to the electrical distribution grid instead of dissipating it as heat.
- *Solid residues.* Any activity that reduces the amount of solid residue in existence justifies bonus scoring. An example would be the use of incoming packaging for the manufacture of a product rather than the use of virgin materials for the same purpose.
- *Liquid residues.* Any activity that renders liquid flows or residues more fit for further service justifies bonus scoring. The purification of incoming water supplies was cited above as an example, but the purification and subsequent further use of any incoming liquid stream would serve as well. Another candidate would be any process that results in the recovery of usable liquid from an existing product of some kind.
- *Gaseous residues.* Any activity that purifies gases and renders them more fit for further service justifies bonus scoring. The removal of pollutants is the most obvious example, and it is commonly done in museums in urban areas in order to protect the artifacts within.

Remanufacture, in particular, provides many opportunities for product, process, and facility improvement. A building may use rainwater that was previously discarded, or a product designer may replace a hazardous material with one that is less hazardous. Bonus scoring is perhaps most reasonable when a rebuilt or refurbished product or facility is being assessed because, in that case, there is great potential to transform materials or environments whose use has not been exemplary and thus create improvements.

The system of bonus scoring is, in some sense, a way of capturing within the SLCA information that might be omitted in a conventional streamlined LCA but captured in a comprehensive LCA. Consider a process in which power is generated and applied to the electrical distribution grid. In an LCA, this activity would be reflected in lower impacts associated with the principal manufactured product because electricity would become a co-product and some of the process burdens would be allocated to it. In the SLCA with bonus scoring, this benefit is captured, and the bonus system thus constitutes a means of performing a rough allocation of process burdens and benefits.

13.3 INTRODUCING LOCALIZATION INTO SCLA MATRIX TECHNIQUES

In the discussion of impact analysis in Chapter 4, the concept of *localization* was introduced, in which the data from preceding LCA steps are related to the particular circumstances in which environmental impacts occur: the temporal scale, the spatial scale, and the location or area that is the receptor for the impact. The difficulty of quantifying these factors has heretofore made it difficult or impossible to include localization in a comprehensive LCA. It is possible, however, to simplify the localization concept and draw upon the approaches discussed in earlier chapters to localize SLCA ratings, as shown below.

To begin, recall that it was mentioned in Chapter 1 that the U.S. EPA's Science Advisory Board uses several ranking guidelines to sort environmental impacts into priority order:

- The penalty for being wrong.
- The spatial scale of the impact.
- The severity of the hazard.
- The degree of exposure.

Rewording these guidelines to make them more comprehensive and less anthropocentric gives the following as suitable parameters for localization:

- *Time*. This parameter considers the time scale of the penalty for being wrong (longer remediation or reversibility times being of more concern than shorter times).
- *Distance*. This parameter considers the spatial scale of the impact (large distances being worse than small).
- *Peril*. This parameter considers the potential severity of the hazard (more toxic substances being of more concern than less toxic substances, more sensitive ecosystems being of more concern than less sensitive, complete species loss being of more concern than partial loss, etc.).
- *Exposure*. This parameter considers the magnitude and degree of exposure (the exposure of larger populations being of more concern than small populations, well-sequestered substances being of less concern than readily mobilized substances, etc.).

How might these parameters be incorporated into the streamlined assessment structure that has been outlined in earlier chapters? It seems reasonable to begin with the 5×5 matrix approach and see if it can be appropriately enhanced. To do so, suppose that the four parameters are treated as an additive function of the matrix entries, and that they are given equal weight. (Obviously, a number of alternative formulations of the parametric function might be considered.) As before, each of the localization parameters is permitted only integer values between 0 (the least favorable rating) and 4 (the most favorable). Thus, the localized matrix element values are given by

196 Chapter 13 Upgraded Streamlining

$$\mathbf{M}_{i,j} = \mathbf{M}_{i,j} \frac{t_{i,j} + d_{i,j} + p_{i,j} + e_{i,j}}{16} \tag{13.1}$$

where $\mathbf{M}_{i,j}$ is the localized matrix element value,
$\mathbf{M}_{i,j}$ is the unlocalized matrix element value,
$t_{i,j}$ is the time scale over which the stress acts,
$d_{i,j}$ is the spatial scale over which the stress acts,
$p_{i,j}$ is the degree of peril attributable to the stress, and
$e_{i,j}$ is the degree of exposure.

The result is that the matrix element values are localized to take circumstantial differences into account.

The next step is to specify how to evaluate the localization parameters. What are favorable and unfavorable temporal and spatial scales? A few of the more common environmental concerns are located in time-distance space in Figure 13.1. In accordance with the guidelines, the concerns near the upper-right corner should be regarded as potentially more damaging than those in the lower left, while those with high position on one of the two axes should be rated intermediate in some way.

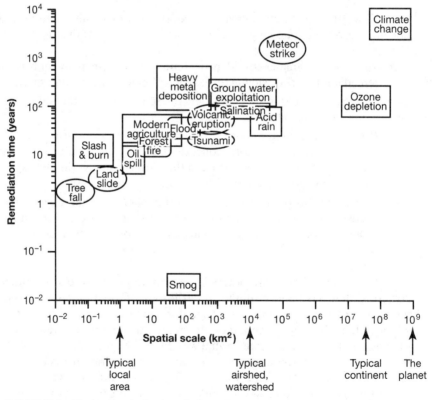

FIGURE 13.1 The location in the time-distance plane of several common environmental concerns. (Adapted from A.P. Dobson, A.D. Bradshaw, and A.J.M. Baker, Hopes for the future: Restoration ecology and conservation biology, *Science*, 277 (1997) 515.)

Thus, for the time and distance scales, the values of the parameters are defined as follows:

Time and Distance Characteristics Related to Parametric Values

	Parametric Value				
	0	1	2	3	4
t	$> 10^3$ days	$10^2 - 10^3$ days	$10^1 - 10^2$ days	$1 - 10$ days	< 1 days
d	$> 10^3$ km	$10^2 - 10^3$ km	$10^1 - 10^2$ km	$1 - 10$ km	< 1 km

Thus, industrially related environmental stressors that have a very short time scale have high parametric value, those with long time scales have lower values. Local effects are assigned d = 3 or 4, regional effects d = 1 or 2, global effects d = 0.

Specifying the values for peril and exposure is more difficult, because there are many different receptors that must be addressed. For impacts on resource consumption, for example, the values can be specified as the following:

Peril and Exposure Characteristics Related to Resource Consumption

	Parametric Value				
	0	1	2	3	4
p	Highly limited global supply	Limited global supply	Adequate global supply	Abundant global supply	Unlimited global supply
e	Highly limited accessible supply	Limited accessible supply	Adequate accessible supply	Abundant accessible supply	Unlimited accessible supply

where "accessible" indicates the abundance of the resource within or readily available to the national unit where the localization is being applied. On a global scale, the classes of supply for the elements are given in Table 13.1. For resources whose global supply is scarce or very scarce, t = 0 should be assigned because the time impact of the use of those resources is essentially infinite. Similarly, d = 0 is appropriate for those resources.

In the case of biodiversity, the guidelines are as follows:

TABLE 13.1 Global Supply Status of the Elements

Unlimited supply	A, Br, Ca, Cl, Kr, Mg, N, Na, Ne, O, Rn, Si, Xe
Abundant supply	Al(Ga), C, Fe, H, K, S, Ti
Adequate supply	I, Li, P, Rb, Sr
Limited supply	Co, Cr, Mo(Rh), Ni, Pb(As,Bi), Pt(Ir, Os, Pa, Rh, Rn), Zr(Hf)
Highly limited supply	Ag, Au, Cu(As, Se, Te), He, Hg, Sn, Zn(Cd, Ge, In, Th)

In the case of chromium and the platinum metals, supply is adequate, but virtually all is from South Africa and Zimbabwe. This geographical distribution makes supplies potentially subject to cartel control. For cobalt and nickel, maintenance of supplies will require mining seafloor nodules. Elements in parentheses are "hitchhikers" normally mined along with the indicated "parent element".

Peril and Exposure Characteristics Related to Biodiversity

	Parametric Value				
	0	1	2	3	4
p	Complete ecosystem loss	Extensive ecosystem loss	Modest ecosystem loss	Slight ecosystem loss	No ecosystem loss
e	$> 10^3$ ha	$10^2 - 10^3$ ha	$10 - 10^2$ ha	$1 - 10$ ha	< 1 ha

Evaluating these parameters follows directly from the size and type of habitat destruction envisioned.

For energy use, the parameter values relate to the power source used to generate the energy, both on a global and a local basis.

Peril and Exposure Characteristics Related to Energy Use

	Parametric Value				
	0	1	2	3	4
p	Globally very scarce energy source	Globally scarce energy source	Globally abundant non-renewable energy source	Renewable energy source	Benign energy source
e	Very scarce accessible energy source	Scarce accessible energy source	Abundant accessible energy source	Renewable accessible energy source	Benign accessible energy source

where "accessible" indicates the abundance of the energy resource within or readily available to the national unit where the localization is being applied.

No energy sources are currently deemed "very scarce." "Scarce" energy sources are petroleum and natural gas (at least from the perspective of a century or two). "Abundant" energy sources are coal and uranium. "Renewable" energy sources are biomass, water power, wind power, and solar power.

The local and regional water supply situation is a special case of resource availability. Unlike minerals or petroleum, water is not shipped around the world from locations where it is abundant to those where it is not, so the water supply in the immediate area, and the degree of use of that supply, are a topic for careful assessment. The guidelines are as follows:

Peril and Exposure Characteristics Related to Water Use

	Parametric Value				
	0	1	2	3	4
p	Accessible water very scarce	Accessible water scarce	Accessible water adequate	Accessible water abundant	Accessible water unlimited
e	$> 10^5$ people	$10^4 - 10^5$ people	$10^3 - 10^4$ people	$10^2 - 10^3$ people	$< 10^2$ people

Similarly, we can consider solid residues, how their end of life sequence is likely to be carried out, and how many people are in the immediate locale.

Peril and Exposure Characteristics Related to Solid Residue Generation

	Parametric Value				
	0	1	2	3	4
p	Very scarce local recycling and disposal facilities	Scarce local recycling and disposal facilities	Adequate local recycling and disposal facilities	Abundant local recycling and disposal facilities	Ubiquitous local recycling and disposal facilities
e	$> 10^5$ people	$10^4 - 10^5$ people	$10^3 - 10^4$ people	$10^2 - 10^3$ people	$< 10^2$ people

The distance scale for solid residues is small, typically the dimensions of the immediate area of the landfill or incinerator. The time scale is short for incineration, long for landfills.

In the case of liquids, the peril refers to toxicity and the exposure to the geographical surface area potentially affected. The local and regional scales differ only in the magnitude of the affected areas.

Peril and Exposure Characteristics Related to Liquid Residue Generation: Local Impacts

	Parametric Value				
	0	1	2	3	4
p	Highly hazardous; water qual. poor	Weakly hazardous; water qual. poor	Highly hazardous; water qual. good	Weakly hazardous; water qual. good	Non-hazardous
e	$> 10^3$ ha	$10^2 - 10^3$ ha	$10 - 10^2$ ha	$1 - 10$ ha	< 1 ha

Peril and Exposure Characteristics Related to Liquid Residue Generation: Regional Impacts

	Parametric Value				
	0	1	2	3	4
p	Highly hazardous; water qual. poor	Weakly hazardous; water qual. poor	Highly hazardous; water qual. good	Weakly hazardous; water qual. good	Non-hazardous
e	$> 10^5$ ha	$10^4 - 10^5$ ha	$10^3 - 10^4$ ha	$10^2 - 10^3$ ha	$< 10^2$ ha

Gases can place stresses on ecosystems on three distance scales: local, regional, and global. Those that should be judged on the local scale include emittants that are inherently toxic, such as carbon monoxide, polynuclear aromatic hydrocarbons and their nitrated analogues, and industrial gases with low exposure limits. Also included in the local group are gases that are precursors of photochemical smog: volatile organic hydrocarbons (VOCs) and oxides of nitrogen (NO_x).

Peril and Exposure Characteristics Related to Gaseous Residue Generation: Local Impacts

	Parametric Value				
	0	1	2	3	4
p	Highly hazardous; water qual. poor	Weakly hazardous; water qual. poor	Highly hazardous; water qual. good	Weakly hazardous; water qual. good	Non-hazardous
e	$> 10^5$ people	$10^4 - 10^5$ people	$10^3 - 10^4$ people	$10^2 - 10^3$ people	$< 10^2$ people

For emissions leading to acid precipitation, i.e., SO_2, NO_x, HCl, or organic acids, the following applies:

Peril and Exposure Characteristics Related to Gaseous Residue Generation: Regional Impacts

	Parametric Value				
	0	1	2	3	4
p	Essentially unbuffered	Poorly buffered	Moderately buffered	Well buffered	Extensively buffered
e	$> 10^7$ ha	$10^6 - 10^7$ ha	$10^5 - 10^6$ ha	$10^4 - 10^5$ ha	$< 10^4$ ha

where the peril refers to the buffering capacity of the receptor watersheds, and the exposure to the geographical area of the watersheds. Information concerning these data are generally available from environmental scientists.

Peril and Exposure Characteristics Related to Gaseous Residue Generation: Global Impacts

	Parametric Value				
	0	1	2	3	4
p	Very severe climate or ozone impact	Severe climate or ozone impact	Moderate climate or ozone impact	Small climate or ozone impact	No climate or ozone impact
e	*	*	*	*	*

e is always zero because the entire world's population is exposed to global impacts.

In the case of global impacts, the following definitions apply: "Very severe" emittants are CFC-11, CFC-12, and CFC-113. "Severe" emittants are CH_4, N_2O, and CO_2. "Moderate" emittants are HCFCs. "Small" emittants are CO and HFCs. "No effect" emittants are VOCs and NO_x.

Section 13.3 Introducing Localization into SCLA Matrix Techniques 201

The localization parameters for the final life stage relate to the capabilities for the recycling and reuse of materials.

Peril and Exposure Characteristics Related to Materials Recycling

	Parametric Value				
	0	1	2	3	4
p	No available recycling facilities	Sparse and poorly controlled recycling facilities	Adequate recycling facilities	Abundant recycling facilities	Ubiquitous recycling facilities
e	$> 10^5$ people	$10^4 - 10^5$ people	$10^3 - 10^4$ people	$10^2 - 10^3$ people	$< 10^2$ people

How does the localization work in practice? Let us apply the technique to the generic 1950s automobile assessment matrix of Figure 7.5a, where we assume a manufacturing facility in the U.S. Upper Midwest (approximately the states of Illinois, Michigan, Ohio, and Wisconsin). The first step is to generate a list of the characteristics of the geographical area for which localization is to be performed; a typical list is given in Table 13.2. Using that information, together with the guidance tables above, the values of the localization parameters are then derived. For each matrix element, one first determines the stressor that will generate the predominant impact. This step requires some experience and a thorough reading of the information and appendices given in this book, and assumes that the qualitative inventory stage of the SLCA was done thoroughly. Given the identity of that stressor, one then can relate the stressor to the receptor group—ecosystem, global atmosphere, etc.—where the impact will occur, and the tables above then give the parameter values. The localized matrix element values

TABLE 13.2 Assumed Manufacturing and Operating Site Characteristics for Generic 1950s Automobiles

Local air quality	Generally good, but smog and particulate matter are from time to time higher than desirable.
Local and regional water	Poor quality but abundant.
Solid residues	Disposal facilities abundant, no recycling.
Energy supply for electricity	70% coal; 20% petroleum; 10% nuclear.
Energy supply for automobiles	Gasoline.
Regional soil buffering against acidity	Well-buffered.
Local energy sources	Abundant coal, scarce petroleum, scarce natural gas.
Local metal, mineral sources	Abundant iron, other resources scarce.
Materials recycling	Limited to modest iron recycling.
Land area in which biodiversity impacts occur	40 ha (one manufacturing plant).
Land area in which local liquid residue impacts occur	80 ha (one small lake).
Land area in which regional liquid residue impacts occur	4.7×10^5 ha (1/10 of average area of Lakes Erie, Huron, and Michigan).
Land area in which regional gaseous residue impacts occur	1.1×10^7 ha (size of state of Ohio).
People affected by solid residue generation	5×10^3 (one town near a landfill).
People affected by local gaseous residue generation	1.5×10^6 (one very large city).

TABLE 13.3 Time and Distance Parametric Values for Generic 1950s Automobiles in the Upper Midwest

Matrix Element	Receptor Group	Predominant Stressor	$t_{i,j}$	$d_{i,j}$	$P_{i,j}$	$e_{i,j}$	$M_{i,j}$	$\mathbf{M}_{i,j}$
1,1	RC	Steel	1	2	1	2	2	0.8
1.2	E	Coal	0	0	2	2	2	0.5
1,3	SR	Ore	1	3	1	2	3	1.3
1,4	LR	Oil	2	4	1	0	3	1.3
1,5	GR	SO_2	2	1	3	0	2	0.8
2,1	GG	CFC-113	0	0	0	0	0	0
2,2	E	Coal	0	0	2	2	1	0.3
2,3	SR	Scrap	1	3	1	2	2	0.9
2,4	LR	Solvents	2	3	1	0	2	0.8
2,5	GR	VOC	3	2	4	0	1	0.6
3,1	RC	Packaging	2	3	3	3	3	2.1
3,2	E	Coal	0	0	2	2	2	0.5
3,3	SR	Packaging	2	3	1	2	3	1.5
3,4	LL	Oil	4	4	1	4	4	3.3
3,5	GR	CO_2	0	0	1	0	2	0.6
4,1	RC	Oil	0	0	1	1	1	0.1
4,2	E	CO_2	0	0	1	0	0	0
4,3	SR	Tires	1	3	1	2	1	0.4
4,4	LL	Oil	2	4	1	4	1	0.7
4,5	GR	NO_x, VOC	3	2	3	0	0	0
5,1	RC	Steel	1	2	1	2	3	1.1
5,2	E	Coal	0	0	2	2	2	0.5
5,3	MR	Steel	1	2	1	2	2	0.8
5,4	LL	Oil	2	4	1	4	3	2.1
5,5	GL	VOC	3	2	1	0	1	0.4

Receptor groups are defined as follows:
RC, resource consumption
B, biodiversity
E, energy use
W, water use
SR, solid residue generation
LL, liquid residue generation, local impacts
LR, liquid residue generation, regional impacts
GL, gaseous residue generation, local impacts
GR, gaseous residue generation, regional impacts
GG, gaseous residue generation, global impacts
MR, materials recycling.

are then calculated using Equation 13.1. Table 13.3 presents those values for the present exercise, together with a brief explanation of the controlling inventory analysis item responsible for the rating.

The ensemble localized matrix values, to the nearest tenth of a unit, are given in Table 13.4. It is interesting to compare the values in this table with those in Figure 7.5, where the unlocalized assessment was presented. In that case, the life stage with the lowest rating was customer use, with manufacturing the next lowest stage. That rating holds true after localization as well. In the case of environmental stressors, however, the

TABLE 13.4 The Localized Assessment Matrix for the Generic 1950s Automobile

Life Stage	Environmental Stressor					Total
	Materials Choice	Energy Use	Solid Residues	Liquid Residues	Gaseous Residues	
Premanufacture	0.8	0.5	1.3	1.3	0.8	4.7/20
Product Manufacture	0	0.3	0.9	0.8	0.6	2.6/20
Product Delivery	2.1	0.5	1.5	3.3	0.1	7.5/20
Product Use	0.1	0	0.4	0.7	0	1.2/20
Refurbishment, Recycling, Disposal	1.1	0.5	0.8	2.1	0.4	4.9/20
Total	4.1/20	1.8/20	4.9/20	8.2/20	1.9/20	20.9/100

order shifts: that of most concern in the unlocalized case was gaseous emissions, with energy use next. After localization, energy use is the lowest, reflecting the lack of any significant source of petroleum in the region for which localization is being conducted.

Examination of some of the other low-scoring matrix elements, and the localization factors leading to those scores, reveals some general attributes of localization:

- If materials that must be acquired from distant sources are used, the localization score for the product is degraded, especially if those sources of materials may be unreliable due to political factors.
- Poor local water quality results in reasonably high impact scores for even small oil and solvent residues.
- Poor local air quality results in reasonably high impact scores, even for modest VOC and NO_x residues.
- The lack of good recycling systems nullifies good choices of recyclable materials, as well as conscientious design for recyclability efforts.
- Restricted local water supplies put a high premium on the use of even modest amounts of water.
- Energy use scoring is independent of time and space, because its most important impacts are global, but the score is heavily influenced by the availability of adequate local sources of energy.

The summed matrix element value for the localized matrix is about 50 percent below the score for the unlocalized matrix. It is a general rule that localization will always produce a lower score, not a higher one, because local factors render certain impacts more serious than they might be if regarded solely on a global basis. The local factors are closely tied to governmental policies on such things as pollution control, infrastructure, and industrial development. Localization is in some sense, therefore, the recognition of political factors within the LCA/SLCA activity.

13.4 WEIGHTING THE MATRICES BY CONSENSUS

The localized SLCA approach outlined above has surmounted many of the objections to traditional LCA/SLCA techniques by incorporating in a semiquantitative way the spatial and temporal scales of the environmental impacts and the degree of exposure. Societal valuation is not implicitly included, however, because similar numerical ratings will be produced by all impacts with similar spatial and temporal scales, perils, and exposures. This result is a consequence of the procedure, in which the localization parameter values have been technologically assigned. To include societal valuation, it is necessary to find a way to reflect valuation in the assessment, and in the beginning this step can be considered independently of localization; they will be integrated at the end of this chapter.

An important perspective is that, although much useful information is provided by the simple matrices and target plots, localized or not, there is an underlying uneasiness in that equal importance has been given to, for example, the "product delivery" and "in-service" life stages. Because the latter very often has a greater environmental effect, it is appropriate to contemplate some form of matrix element weighting to reflect this information. Thus, instead of the initial approach, in which each matrix element had a maximum value of 4 and the overall assessment value was given by summing the matrix element values, a set of weighting factors $\omega_{i,j}$ is sought that reflects differences in life-stage impact so that each matrix element will have a value given by $\omega_{i,j} M_{i,j}$, and for which the overall rating is given by

$$R_{ERP} = \sum_i \sum_j \omega_{i,j} M_{i,j} \qquad (13.2)$$

Determining in a comprehensive way the appropriate life-stage weightings requires a complete life-cycle analysis, and is in conflict with the desire to streamline the LCA process. The assessment can be intuitively improved, however, if the assessment team chooses the life stage likely to produce the most severe environmental impacts and arbitrarily weights that life stage as, say, one-half the total assessment value, with the other four life stages arbitrarily weighted at, say, one-eighth the total value. For the automobile example that has been used previously, where the in-service life stage almost certainly results in the most severe of the environmental impacts, the in-service stage is chosen as the dominant one for halftile-octile weighting. The weighting factors are then those shown in Table 13.5.

If the matrix element values are recomputed using the weighting factors, the resultant evaluation for the generic 1950s automobile is that shown in Table 13.6, where the increase in influence of life stage 4 is evident and where the overall product rating has dropped from the 44 of Figure 7.5 to 34. This result can be made even more apparent by the target plot. However, the construction of the target plot now requires an additional step in order to demonstrate the degree of departure from optimum of a matrix element value. The procedure is to compute the "deficit matrix," the difference between perfect matrix element scores and actual scores, given by

$$\delta_{i,j} = [\omega_{i,j} M_{i,j}]_{max} - \omega_{i,j} M_{i,j} \qquad (13.3)$$

For the 1950s automobile example, the resulting target plot is shown in Figure 13.2 where, in comparison with Figure 7.6, the deficiencies in environmental responsibility during the in-service life stage are quite obvious.

TABLE 13.5 Weighting Factors for Singly- and Doubly-Weighted Matrices

The matrix of weighting factors for a dominant life stage:

$$\omega_{i,j} = \begin{bmatrix} 0.625 & 0.625 & 0.625 & 0.625 & 0.625 \\ 0.625 & 0.625 & 0.625 & 0.625 & 0.625 \\ 0.625 & 0.625 & 0.625 & 0.625 & 0.625 \\ 2.5 & 2.5 & 2.5 & 2.5 & 2.5 \\ 0.625 & 0.625 & 0.625 & 0.625 & 0.625 \end{bmatrix}$$

The matrix of weighting factors for a dominant environmentally-related attribute:

$$\phi_{i,j} = \begin{bmatrix} 0.625 & 2.5 & 0.625 & 0.625 & 0.625 \\ 0.625 & 2.5 & 0.625 & 0.625 & 0.625 \\ 0.625 & 2.5 & 0.625 & 0.625 & 0.625 \\ 0.625 & 2.5 & 0.625 & 0.625 & 0.625 \\ 0.625 & 2.5 & 0.625 & 0.625 & 0.625 \end{bmatrix}$$

The matrix of weighting factors for a doubly-weighted matrix:

$$\omega_{i,j}\,\phi_{i,j} = \begin{bmatrix} 0.39 & 1.56 & 0.39 & 0.39 & 0.39 \\ 0.39 & 1.56 & 0.39 & 0.39 & 0.39 \\ 0.39 & 1.56 & 0.39 & 0.39 & 0.39 \\ 1.56 & 6.25 & 1.56 & 1.56 & 1.56 \\ 0.39 & 1.56 & 0.39 & 0.39 & 0.39 \end{bmatrix}$$

TABLE 13.6 The Singly-Weighted Environmentally Responsible Product Assessment for the Generic 1950s Automobile

Life Stage	Environmental Stressor					
	Materials Choice	Energy Use	Solid Residues	Liquid Residues	Gaseous Residues	Total
Premanufacture	1.25	1.25	1.88	1.88	1.25	7.5/12.5
Product manufacture	0	0.63	1.25	1.25	0.63	3.8/12.5
Product delivery	1.88	1.25	1.88	2.50	1.25	8.8/12.5
Product use	2.50	0	2.50	2.50	0	7.5/50
Refurbishment, recycling, disposal	1.88	1.25	1.25	1.88	0.63	6.8/12.5
Total	7.5/20	4.4/20	8.7/20	10.0/20	3.8/20	34.4/100

Although the in-service life stage will be the one most highly weighted for many products, it is not difficult to identify products for which other life stages can be assumed to dominate the environmental impacts. Examples are given in Table 13.7.

Just as some life stages produce larger environmental impacts than others, some environmental impacts are of more concern than others. For a specific product, the priority stressor is thus characterized by some combination of the magnitude of the stressor's impacts and the degree to which the impacts meet the high-risk criteria.

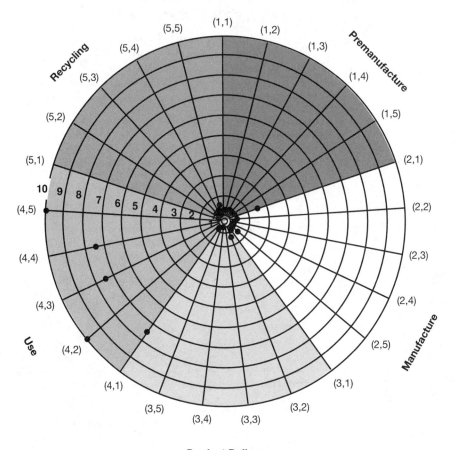

FIGURE 13.2 A target plot for the display of the consensus-weighted matrix assessment results for the generic automobile of the 1950s.

Determining priorities rigorously requires, as before, a comprehensive and defendable life-cycle assessment, but Table 13.7 suggests examples of products for which individual environmental stressors would probably be regarded as dominant by a consensus of experts.

In the case of the 1950s automobile, one could make a case for either energy use or gaseous emissions being the dominant stressor, but for demonstration purposes energy use will be selected here. As with the life-stage weightings, the energy use stressor is arbitrarily assigned half the total assessment value, while the other environmental stressors receive one-eighth each. The second weighting factor $\phi_{i,j}$ is then as shown in Table 13.5; the doubly weighted matrix element values are given by the product $\omega_{i,j} \, \phi_{i,j} \, M_{i,j}$, and the overall rating by

TABLE 13.7 Examples of Products for Which Different Life Stages or Environmentally Related Attributes Probably Dominate

Life Stages
Premanufacture	Laptop computer
Product manufacture	Sheet aluminum from bauxite ore
Product delivery	Traditionally packaged compact disc
Product use	Automobile
Refurbishment, recycling, disposal	Mercury relays

Environmentally-Related Attributes
Materials choice	Medical products with radioactive materials
Energy use	Hair dryers
Solid residues	"Convenience" foods
Liquid residues	Pesticides
Gaseous residues	CFC propellant sprays

TABLE 13.8 The Doubly Weighted Environmentally Responsible Product Assessment for the Generic 1950s Automobile

	Environmental Stressor					
Life Stage	Materials Choice	Energy Use	Solid Residues	Liquid Residues	Gaseous Residues	Total
Premanufacture	0.78	3.13	1.17	1.17	0.78	7.03/12.5
Product manufacture	0	1.56	0.78	0.78	0.39	3.51/12.5
Product delivery	1.17	3.13	1.17	1.56	0.78	7.81/12.5
Product use	1.56	0	1.56	1.56	0	4.68/50
Refurbishment, recycling, disposal	1.17	3.13	0.78	1.17	0.39	6.64/12.5
Total	4.68/12.5	10.95/50	5.46/12.5	6.24/12.5	2.34/12.5	29.67/100

$$R_{ERP} = \sum_i \sum_j \omega_{i,j}\, \phi_{i,j}\, M_{i,j} \qquad (13.4)$$

The matrix result for the 1950s automobile is shown in Table 13.8, the overall rating now having dropped from 34 to 30 and the importance of energy use now being emphasized.

The target plot for the doubly weighted matrix is formed in a manner similar to that for the singly weighted matrix, with the deficits being computed as

$$\Delta_{i,j} = [\omega_{i,j}\, \phi_{i,j}\, M_{i,j}]_{max} - \omega_{i,j}\, \phi_{i,j}\, M_{i,j} \qquad (13.5)$$

The resulting target plot in Figure 13.3 shows dramatically the importance of the in-service life stage, the importance of the energy use environmental stressor, and (most significantly) the failure of the 1950s vehicle to reflect these priorities.

13.5 WEIGHTING THE MATRICES BY THE GRAND OBJECTIVES

Weighting matrices by consensus may be perfectly adequate for many purposes, but the valuation step lacks intellectual rigor. An alternative only slightly more complex is

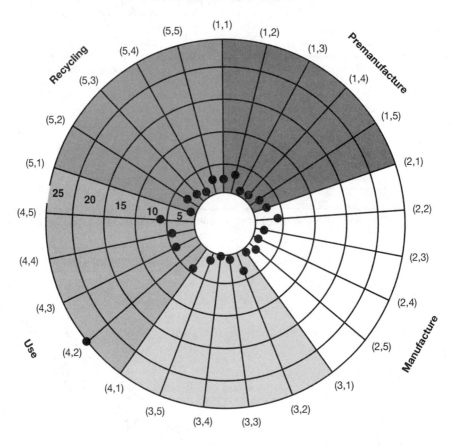

FIGURE 13.3 A target plot for the display of the consensus doubly weighted matrix assessment results for the generic automobile of the 1950s.

to relate the valuation to the grand objectives. Having grouped the environmental concerns by priority in Chapter 1, one next needs to identify the associated activities that should be examined as part of a life-cycle assessment. As Figure 1.3 indicates, this is a step that should be done by environmental scientists and industrial ecologists working in concert, the former to provide perspectives on environmental processes, the latter on industrial processes. A sample of such a list constitutes Table 1.3; it is not intended to be comprehensive, but includes most of the common societal activities having impacts on the environment. Should the assessor wish, it may be made as detailed as needed. In any case, it serves as a guide for the matrix scoring operation.

If no valuation is being incorporated (i.e., if environmental concerns are not grouped or otherwise prioritized), the process of assigning a number to a matrix element follows the procedure given in the top of Table 13.9, where, for illustration,

TABLE 13.9 Scoring Guidance for Matrix Element 2,5: Manufacturing—Gaseous Residues

If the following applies, the matrix element rating is 0:
Gaseous manufacturing residues are very large in quantity relative to similar facilities elsewhere and no reuse/recycling programs have been implemented.
If the following applies, the matrix element rating is 4:
Gaseous manufacturing residues are negligible.

Valuation Not Considered
If neither a rating of 4 or 0 is assigned, review the emittants, both species and quantities, and their relationships to environmental concerns.
Assign a rating of 3 if gaseous emissions are very small in quantity relative to similar facilities elsewhere and none of the emittants is sensitive from a regulatory standpoint.
Assign a rating of 2 if gaseous emissions are moderate in quantity relative to similar facilities elsewhere and/or are moderately sensitive from a regulatory standpoint.
Assign a rating of 1 if gaseous emissions are large in quantity relative to similar facilities elsewhere and/or highly sensitive from a regulatory standpoint.

Valuation Explicitly Considered
If neither a rating of 4 or 0 is assigned, review the emittants, both species and quantities, and their relationships to environmental concerns.
Assign a rating of 3 if gaseous emissions have negligible impacts on crucial concerns and low to moderate impacts on other concerns.
Assign a rating of 2 if gaseous emissions have moderate impacts on crucial concerns and/or high impacts on other concerns.
Assign a rating of 1 if gaseous emissions have high impacts on crucial concerns and/or very high impacts on other concerns.

assume that an industrial product is being assessed and that the value for matrix element 2,5 (manufacturing, gaseous emissions) is being determined. (Note that unlocalized values are discussed here.) For the extreme ratings of 0 or 4, the evaluation in this case, at least) is independent of whether or not valuation is being incorporated. For the intermediate ratings, however, the guidance is seen to rely heavily on the relative quantity of emissions and on the potential for regulatory action as a consequence. If regulatory potential were indeed an accurate reflection of prioritized environmental impacts, such a procedure would be perfectly satisfactory. In practice, however, regulatory structures are well known to emphasize anthropocentric, local, current, and well-publicized impacts and to de-emphasize or ignore nonanthropocentric, nonlocal, more subtle, and future impacts. Consequently, recommendations arising from an assessment leaning on regulatory guidance tend to have a short-term focus in time, space, and law, and generally ignore or operate counter to the perspectives and concerns of the environmental science community.

Conversely, if a grouped prioritization approach is being taken, the process of assigning a value to matrix element 2,5 occurs in four steps:

Step 1 Identify the gases emitted from the manufacturing facility.
Step 2 Measure or estimate the emission fluxes of each of the gases.
Step 3 Relate the emissions to the associated environmental concerns.
Step 4 Take the results of Step 3 and assign a matrix element rating.

TABLE 13.10 Relating Environmental Concerns to Gaseous Causative Agents

Concern	Gaseous Causative Agents
Climate change	CO_2, CH_4, CFCs, N_2O
Biodiversity	Toxins, carcinogens, mutagens
Ozone depletion	CFCs, HCFCs, Halons, N_2O
Human organism damage	PAH, other toxins, carcinogens, mutagens
Water quality	None
Fossil fuel depletion	None
Soil depletion	None
Suboptimal land use	None
Acid deposition	SO_2, NO_x, HCl
Smog	NO_x, VOCs, particulate matter
Aesthetics	Particulate matter, SO_2
Non-fossil fuel depletion	None
Oil spills	None
Radioactive gases	Radionuclides
Odor	H_2S, mercaptans, amines, etc.
Thermal pollution	None
Landfill exhaustion	None

Steps 1 and 2 are self-evident. For Step 3, the relationships are given in Table 13.10. (In the compilation of such a table, environmental scientists must be consulted.) Step 4 is accomplished as in Table 13.9. Thus, if a facility were emitting large amounts of CFCs from metal cleaning operations, a value of 0 would be assigned. If the facility was emitting modest amounts of VOCs and small amounts of H_2S, a value of 2 might be assigned. ("Large," "modest," and "small" may be operationally defined with respect to the emissions of similar facilities.) Corporate checklists and guidelines, as in the Appendices, provide guidance in this operation.

The goal of a product life-cycle assessment is to produce a series of recommendations for the product design team. These recommendations should be designed to improve the product design and thus minimize the product's environmental impact. To demonstrate, imagine that an LCA or SLCA assessment of steel office-equipment frames generates the recommendations shown in Table 13.11. Notice that each specific recommendation in Table 13.11 carries with it a three-element number, which identifies the grand objective, the environmental concern, and the targeted activity, and thus directly connects an action within a design or manufacturing team to the grand objective that will be influenced. If assessors ensure that no recommendation is ever given without such a number attached, the result will be a guarantee that the recommendation will truly contribute to ameliorating important environmental impacts. The identification of the crucial concerns by the use of bold font serves as an aid in prioritizing the recommendations for the use of design, development, and manufacturing engineers.

The approach described above obviously relates the assessment of a particular matrix element (i.e., a single life stage and a single environmental stress) more closely to the crucial environmental concerns than if prioritization of environmental concerns is ignored. Such a procedure, while intellectually more satisfying, may still have little

Section 13.5   Weighting the Matrices by the Grand Objectives   211

TABLE 13.11   Design for Environment Recommendations for Office Equipment Frames of Painted Steel[‡]

- Do not use CFCs for metal cleaning (Ω_3, 3.1, Ω_1, **1.8**)
- Do not use chromate anticorrosive coatings (Ω_3, 2.4, Ω_1, **4.9**)
- Design the frames to facilitate disassembly and recovery of materials at end of life (Ω_1, 1.1, Ω_1, 6.1, Ω_4, 9.7, Ω_2, 10.4)
- Use recycled material to the fullest extent possible rather than using virgin material (Ω_1, 1.1, Ω_1, 6.1, Ω_4, 9.7, Ω_2, 10.4)
- Use powder paints rather than oil-based or aqueous-based spray paints (Ω_4, 8.2)
- Minimize the amount of packaging used for the products (Ω_2, 10.4)
- Put recycling instructions on the product packaging (Ω_2, 10.4)

[‡]The numbers refer to the grand objective, the environmental concern, and the specific societal activity to which the recommendation responds. Numbers given in bold face refer to crucial environmental concerns.

Life stage	Environmental Stressor				
	Materials choice	Energy use	Solid residues	Liquid residues	Gaseous residues
Premanufacture		F,C			
Product manufacture		F,C		B,H;W	H,O
Product delivery		F,C			
Product use		F,C		B,H	O
Refurbishment, recycling, disposal	F,C	F,C	F,C		

FIGURE 13.4   The relationship of selected matrix elements to crucial environmental concerns. B = loss of biodiversity; C = global climate change; F = depletion of fossil fuel resources; H = human organism damage; O = stratospheric ozone depletion; W = water availability and quality. The cross-hatching indicates a matrix element in which activity always has an impact on a crucial environmental concern; lined hatching indicates a matrix element in which activity sometimes has an impact on a crucial environmental concern.

effect on the actual scoring, which requires employing the weighting factors of Equation 13.2. From the perspective of the present work, the assignment of weighting factors can be seen to reflect the explicit prioritized grouping of environmental concerns. To begin defining the weighting factors, examine which matrix elements relate to those concerns identified as crucial. Figure 13.4 demonstrates that there are 11 such matrix elements, six indicated by cross-hatching to indicate that any activity of the type indicated will always produce a stress on a crucial environmental concern, and five called out by lined hatching, indicating that the activity will sometimes cause stress. Energy use in all life stages is included in these 11 selected matrix elements, because energy use has impacts on both the fossil fuel resource base and on global climate change. Materials choice and the type, volume, and properties of solid residues at the reuse/recycling life stage appear as well, because each influences whether materials will be reused (an energy-efficient thing to do). Liquid and gaseous residues in manufacturing are included because they potentially relate to organism damage, water use, and ozone depletion; they are also called out in the product use stage, where some of the same stresses may occur.

One could attempt to set weighting factors in very intricate and complex ways, but simple procedures will probably suffice to furnish advice to product designers. A reasonable approach is that a matrix element where activity always creates a stress on a crucial environmental concern receives three times the weight of a matrix element where such stress is rarely or never caused, and that matrix elements where activity sometimes creates such a stress receive twice the weight. If the three weights are designated as ω_H, ω_M, and ω_L (the subscripts indicating high, medium, and low weighting), and 100 is retained as the maximum summed score, the value of the weighting factors is given by

$$\omega_H \cdot 6.4 + \omega_M \cdot 5.4 + \omega_L \cdot 14.4 = 100 \tag{13.6}$$

or

$$3\,\omega_L \cdot 6.4 + 2\,\omega_L \cdot 5.4 + \omega_L \cdot 14.4 = 100 \tag{13.7}$$

and thus, approximately,

$$\omega_H = 1.80;\ \omega_M = 1.20;\ \omega_L = 0.60 \tag{13.8}$$

The difference in justification for recommendations made with and without valuation can readily be demonstrated. For example, a common recommendation for products is to use recycled materials. The justification is that such use minimizes the depletion of resources, and such a justification would be implied by traditional LCA/SLCAs. However, as seen from Table 1.2, depletion of non-fossil fuel resources is not considered a crucial environmental concern. Nonetheless, it is, in general, the case that recycling or reuse of materials that have already been through the extraction, refining, and use stages requires much less energy than the extraction of new virgin materials. Hence, the recommendation for use of recycled materials can be justified on the basis of fossil fuel resources and global climate change concerns.

Thus, the result is the specific inclusion of societally weighted environmental concerns in the LCA/SLCA process. The particular LCA/SLCA technique is not central to the use of grand objectives as a societal driver of industrial decisions; rather, the concept may be incorporated into any suitable LCA/SLCA technique. By defining environmental concerns and the activities related to them, and the design for environment recommendations related in turn to those activities, the environmental approach to technology is linked more directly and obviously to the underlying environmental science and the societal value structure.

Although a precise ranking of prioritization of environmental concerns is inherently contentious and culture- and location-related, several concerns nonetheless are universally regarded as very important, others as less important. The collecting of concerns into importance groups based on the grand objectives informs the LCA/SLCA process without descending into the traditional morass of quantification and that of monetary evaluation. If perhaps not necessarily defendable in detail for policy purposes, the approach described here nonetheless is eminently satisfactory as guidance for product-related decisions by industrial design, development, and manufacturing engineers.

The four-step sequence that relates the grand objectives to the specific recommendations of industrial ecology is not a sequence that can be carried out by a single group of specialists. Rather, three distinct groups of players emerge. The first is the industrialized society itself, which implicitly or explicitly establishes the objectives. The second is the environmental science community, which identifies environmental concerns and the technological and societal activities connected to them. The third is the design for environmental community, which modifies technological activities to minimize their impacts on the crucial environmental concerns. The ensemble of activities of these three groups, together with relevant management, economics, and policy activities, constitutes *industrial ecology*, the "science and technology of sustainability." The goal of sustainability requires the constructive interaction of these groups; it is a goal that has some prospect of success only through the recognition that intellectual consistency and operational progress are essential to successful implementation.

13.6 THE LOCALIZED, VALUATED SLCA

As an example of the application of the full spectrum of SLCA tools, consider once again the generic 1950s automobile. Its localized matrix appears in Table 13.4. To that matrix is now applied the grand objective weighting factors presented in Figure 13.4. The result is the matrix shown in Table 13.12. The overall score has once more decreased somewhat, reflecting the importance that is placed on some of the environmental concerns as related to the grand objectives. So far as relative life stages are concerned, little has changed from either the nonlocalized or localized assessments. In the case of the environmental concerns, however, the column sums show that both the energy use and gaseous residues scores are only about one-tenth that which is possible. However, the grand objectives weighting renders the energy use concern twice as important. The other environmental concerns, with possible scores equal to that for gaseous residues, have scores about 20–30 percent of that which is possible.

TABLE 13.12 The Localized and Grand Objectives-Weighted Assessment for the Generic 1950s Automobile

Life Stage	Environmental Stressor					
	Materials Choice	Energy Use	Solid Residues	Liquid Residues	Gaseous Residues	Total
Premanufacture	0.5	0.9	0.8	0.8	0.5	3.5/16.8
Product manufacture	0	0.5	0.5	1.4	0.7	3.1/24.0
Product delivery	1.3	0.9	0.9	2.0	0.1	5.2/16.8
Product use	0.1	0	0.2	0.8	0	1.1/21.6
Refurbishment, recycling, disposal	1.3	0.9	1.0	1.3	0.2	4.7/21.6
Total	3.2/14.4	3.2/36.0	3.4/14.4	6.3/19.2	1.5/16.8	17.6/100

214 Chapter 13 Upgraded Streamlining

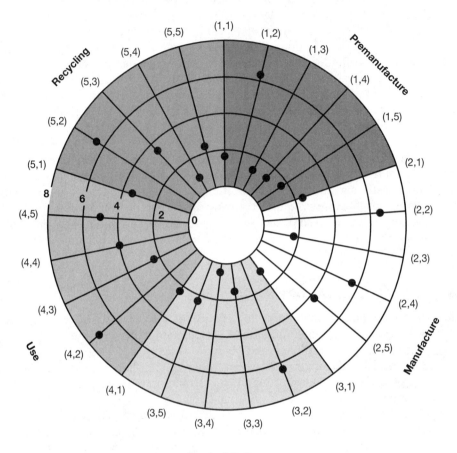

FIGURE 13.5 A target plot for the display of the localized, grand-objectives weighted, assessment matrix results for the generic automobile of the 1950s.

The weighting is again obvious in the target plot, constructed in accordance with Eq. 13.2 and shown in Figure 13.5. As was the case with the simple weighting presented earlier, the energy use stage stands out. This time, however, that position is established not by instinctively driven fact, as was the case earlier, but by reference to the long-range goals of society. The technique used cannot claim perfection; there is too much societal decision making involved to feel that perfection can ever be achieved. The approach is, however, intellectually consistent and promising as guidance to the technologist as well as the citizen.

It is apparent that from the standpoint of the design engineer, there will be little difference in the specific modifications that might be undertaken to lessen the environmental impacts of the product. What has changed, however, is that the actions are now taken in response to the comprehensive use of information related to the characteristics of the location in which the impacts will occur, as well as to societal values. From an intellectual standpoint, every action taken to improve environmental responsibility is now justified in detail.

FURTHER READING

Curran, M.A., Broad-based environmental life cycle assessment, *Environmental Science & Technology, 27* (1993) 430–436.

Graedel, T.E., Weighted matrices as product life-cycle assessment tools, *International Journal of Life Cycle Assessment, 1* (1996) 85–89.

Lindfors, L. G., Summary and recommendations, in *Product Life Cycle Assessment*—Principles and Methodology, Nord 1992:9 (Copenhagen: Nordic Council of Ministers, 1992) pp. 9–23.

EXERCISES

13.1. Discuss the concept of bonus scoring. Is it a useful analytical device from a corporate perspective? From a societal perspective?

13.2. Assume that the localization exercise outlined in Tables 13.2–4 is applied to an automobile manufacturing facility in Norway. All characteristics are the same except that energy is supplied exclusively by hydropower, the water quality in the region is excellent, and the soil buffering is poor. Reassign parametric values and complete a revised assessment matrix.

13.3. Evaluate the utility of choosing and weighting a "most important" life stage or environmental stressor, as described in Section 13.4. What are the benefits and liabilities of this approach?

13.4. A major virtue of SLCAs is their simplicity and efficiency. The process of upgrading SLCAs inevitably introduces more complexity and more requirements for expert judgment. On balance, how much upgrading do you think is optimum? Should a higher level of upgrading be used in some situations but not others?

CHAPTER 14
Reverse LCAs

14.1 INTRODUCTION

Traditional life-cycle assessment generally begins with a product and proceeds to examine the environmental impacts throughout the product's life cycle. An alternative approach is to proceed in reverse: To examine the need that the product is designed to fulfill, to determine the minimal environmental impacts that could be engendered by filling that need, and thereby to design the "perfect green product" for the purpose. This approach, termed *reverse life-cycle assessment* (RLCA), can be demonstrated by examining the environmental impacts attributable to a generic washing machine of current design, and then by reviewing other ways in which the provisioning of clean clothing may be accomplished. RLCA, as used here, is shown to encourage systems thinking and to identify opportunities for innovation in design and in marketing of environmentally responsible products in ways that would be unlikely to arise from a traditional LCA.

This alternative approach to developing a design and then assessing its environmental attributes begins with the broadening of one's perspective and the consideration not of the greenness of a particular design or a particular product, but of the fact that a product exists to fulfill a specific need or desire of society. One may then ask not whether a particular product is green, but rather, from an environmental standpoint, what sort of product is the best way to fill the need or desire that has been identified. This needs-based approach can be studied, in principle, by performing the LCA or SLCA in reverse, that is, by beginning with the environmental characteristics of an ideal product and then working backward to determine the physical design approach that would most nearly satisfy those characteristics.

14.2 LIFE-CYCLE ASSESSMENT OF THE GENERIC WASHING MACHINE

The reverse life-cycle approach can be illustrated by analyzing a familiar consumer appliance: the generic washing machine, which performs 350–400 washes per year and has an average life of 10–14 years. First, the characteristics of a present-day machine will be presented, and then an SLCA analysis will be performed. The results exemplify the attributes, good and bad, of the modern approach to the provisioning of clean clothing.

Section 14.2   Life-Cycle Assessment of the Generic Washing Machine   217

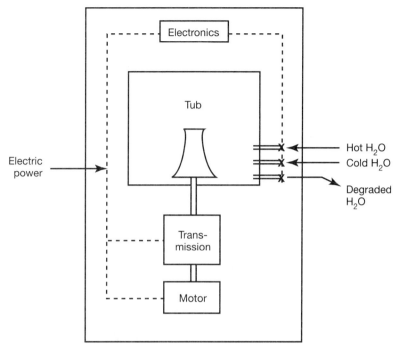

FIGURE 14.1   Components of a generic washing machine.

14.2.1 The Washing Machine Life Cycle

It is useful to begin by reviewing the life cycle of the generic washing machine and the constituent parts and subassemblies from which it is made (Figure 14.1). The life cycle can be summarized in the following stages:

1. Fabrication of individual parts and materials, including the frame, the mechanical components, the electrical and electronic components, and the housing. Much of this fabrication is done by suppliers.
2. System assembly, in which mechanical and electrical components are assembled into frames, plumbing and wiring is added, and system tests are performed.
3. Product delivery, in which finished washing machines are packaged and shipped, sold, and installed on customer premises.
4. In-service use, in which the product performs the washing actions desired by the customer.
5. In-service maintenance, in which repairs and updates to the machine are performed in order to keep it in use.
6. Recycling and/or disposal, in which materials from obsolete systems re-enter materials streams or are lost to the industrial process as a consequence of landfilling, incineration, or other disposal techniques.

218 Chapter 14 Reverse LCAs

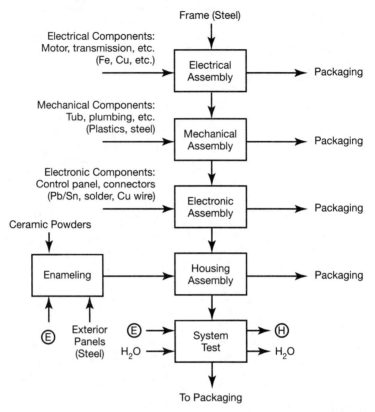

FIGURE 14.2 A life-cycle flow diagram for the generic washing machine, including manufacture, use, maintenance, and end of life. Rectangles indicate process flow steps. Significant energy use is indicated by a circled E, significant heat loss by a circled H.

14.2.2 Assembly Flow

The flow diagram for the washing machine life cycle is shown in Figure 14.2. The sequence proceeds from the bottom of the diagram to the top, components and subassemblies enter from the left of the diagram. Waste products leave processing steps to the left if they are part of a recycling stream, to the right if they are discarded. In general, incoming parts and components are packaged in a variety of plastic and cardboard containers, often with wood bases and metal strapping. No attempt is generally made to achieve uniformity in supplier packaging, or to coordinate or optimize supplier deliveries so as to minimize their environmental impacts.

In the first assembly step, the electromechanical components of the system (motor, transmission, etc.) are affixed in the frame. These components are made largely of steel and copper. Next, the mechanical components are added. These include the plumbing connections, made largely of stainless steel and plastic, and the tub and other components of the washing compartment, most of which are plastic. The third step is the addition of the control panel and of the wiring that connects the panel to the

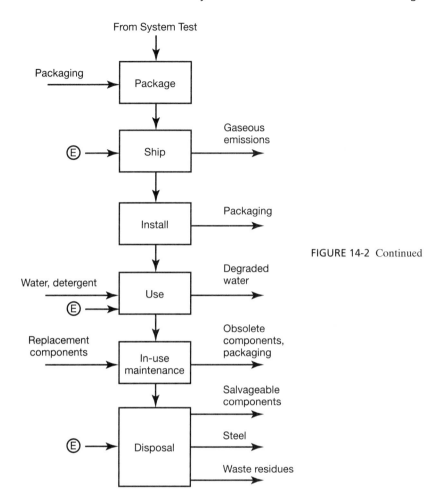

FIGURE 14-2 Continued

electromechanical components. These items incorporate electronic circuit boards as well as copper conductors and plastic cable sheathing. The next step is to affix the external housing: sides, top, front, and back. Most of these items are manufactured of enameled steel. In all four of these stages, the components that are added have arrived at the manufacturing facility in their own packaging; that packaging must be discarded or otherwise dealt with. The final factory step is testing the completed unit, an operation that involves the use of energy and water, but no new materials.

After completion and testing, the washing machine is packaged for shipment to the customer. Typical packaging involves a wood or heavy corrugated cardboard base, some internal polymer cushioning, an external corrugated cardboard box, and steel strapping tape to retain the packaging. Shipment to the point of sale and ultimately to the customer then occurs, normally by truck. The unit is then installed in the home, generally by an employee of the point of sale organization. Disposal or recycling of the packaging is usually left to the customer.

In use, the washing machine consumes energy, heated water, detergent, and perhaps other washing agents in producing clean clothing. During the useful life of the machine, components that become defective are repaired or replaced. Finally, the machine is disposed of, and the appropriate disposal or recycling system deals with the obsolete unit.

14.3 STREAMLINED LCA OF THE GENERIC WASHING MACHINE

14.3.1 The Matrix Approach

As with other products, the DFE characteristics of the generic washing machine can be evaluated with the 5×5 assessment matrix of Chapter 7. The considerations are as follows.

Premanufacture The premanufacture assessment stage could, in principle, treat impacts on the environment as a consequence of the actions needed to extract materials from their natural reservoirs, transport them to processing facilities, purify or separate them by such operations as ore smelting and petroleum refining, utilize the materials in the manufacture of parts and components, and transport the products to the factory where they are incorporated in new washing machines. Such an analysis would include consideration of plastics formulations and treatment of associated industrial waste, air emissions from metal enameling processes, and the like. In practice, such an assessment is generally impractical, and responsible manufacturers deal with the issues by assuring themselves that their suppliers practice environmental responsibility in their business practices. As a result, the premanufacturing assessment stage is restricted to evaluation of the impacts on the *manufacturer's* environmental performance that arise from the materials and design choices by suppliers and their packaging and transport to the manufacturing facility (Table 14.1). Since a washing machine is primarily the result of an assembly of parts supplied by others, the final product's environmental responsibility is in large degree established by the choices made by supplier design engineers.

Component designers, in general, make good materials choices from a DFE standpoint. Few of the materials have any toxicity problems and those materials of some concern are present only in small amounts or in situations where the hazardous aspects are minimized. The use of scarce resources is modest.

Notwithstanding the above, several materials choices should be noted as items to be improved if opportunities present themselves:

- Lead solder on circuit boards. The bioavailability of this lead is not great, and detailed investigations of alternative solders have not yet yielded a suitable

TABLE 14.1 Premanufacture

Element Designation	Element Value and Explanation
Matls. choice (1,1)	0 (Use of lead and chromate.)
Energy use (1,2)	2 (High energy use in deliveries.)
Solid res. (1,3)	2 (Abundant and diverse packaging.)
Liquid res. (1,4)	3 (Mining residues.)
Gas res. (1,5)	2 (Substantial air pollution from deliveries.)

replacement. Nonetheless, the toxicity of lead and its environmental persistence is such that its use anywhere is undesirable if alternatives can be developed.
- Metal frames, bolts, and screws generally have zinc chromate coating for corrosion resistance. The hexavalent chromium in zinc chromate is biotoxic and will eventually enter the soil or water waste flow streams.

A frequent opportunity for improvement at the premanufacture stage is to specify the use of recycled material where possible in items or materials purchased from suppliers. Such an activity is essential for all designers if materials flows are to be sustainable in the long term. Among the possibilities worth considering are recycled metals in equipment frames, recycled plastics as ingredients of plastic parts, and recycled material for boxes and other packaging items. In every case, of course, the recycled material would need to meet appropriate performance specifications and have lower environmental burdens than do the virgin materials it replaces.

Product Manufacture Product manufacture (Table 14.2) is the life stage most likely to have been assessed for environmental impact by industrial engineers, because such aspects of it as emissions to air and water are subject to review by governmental authorities. Other characteristics of manufacture are not covered by such oversight, however, and all facets need to be part of a life-cycle assessment.

In the manufacture of the generic washing machine, the most significant environmental impact is likely to be the enameling operation. In this process, ceramic and glass powders are applied to metal surfaces, which are then heated to some 800°C to melt and fuse the powders to the metal. Large amounts of energy are used, substantial heat is lost, and organic and inorganic gases are produced and emitted. Little solid residue is produced in washing machine manufacture. Most electrical and fluid connections, for example, are supplied in the desired lengths, so no cutting scraps result. Liquid residues are minimal as well. All machines are tested when completed, but the water and detergent are recycled.

Product Delivery This life stage includes the environmental impacts arising from residues generated or energy used during the packaging process, transportation of the finished and packaged product to the customer, product installation, and disposal or recycling of the packaging material (Table 14.3).

The traditional task of the packaging engineer has been to devise packaging that safeguards the product between manufacture and installation. Minimizing cost has been a part of the consideration involved, but the environmental impacts of the packaging have traditionally not been considered. As a consequence, washing machine packaging typically contains a mixture of materials (cardboard, foam, plastic sheet,

TABLE 14.2 Product Manufacture

Element Designation	Element Value and Explanation
Matls. choice (2,1)	2 (Minor lead and powder toxicity concerns.)
Energy use (2,2)	2 (Copious energy use during enameling.)
Solid res. (2,3)	3 (Little solid residue produced.)
Liquid res. (2,4)	3 (Minor liquid residues from testing.)
Gas res. (2,5)	3 (Inorganic and organic emissions from enameling.)

TABLE 14.3 Product Delivery

Element Designation	Element Value and Explanation
Matls. choice (3,1)	1 (Substantial materials diversity in packaging.)
Energy use (3,2)	2 (Energy use in delivery not optimized.)
Solid res. (3,3)	1 (No plans for material recovery and recycling.)
Liquid res. (3,4)	4 (No concerns.)
Gas res. (3,5)	2 (Gaseous emissions in delivery not optimized.)

steel strapping, etc.) and its disposal is left to the customer. Information on proper recycling of the packaging material is seldom provided, so even someone with the intent to recycle does not know, for example, which polymer is used to make the plastic sheet that covers the components to prevent scratches. Properly marking the packaging material obviously does not guarantee that it will be recycled, but improves the probability of that happening.

Environmental Impacts During Customer Use This life stage includes impacts from consumables and maintenance materials that are expended during the period in which the washing machine is used by the customer (Table 14.4). The typical washing machine has significant environmental impacts as a result of in-service operation. The most obvious, and one that deserves the most attention, is the power required to heat the water and for motor and valve operation. Water consumption and discharge are also major issues, especially in regions where water is a scarce resource. The use of detergents and other washing products must be considered, as well as the environmental consequences of the maintenance and repair functions that are inevitably required.

Refurbishment, Recycling, Disposal The fifth life stage assessment includes impacts from product refurbishment and as a consequence of the eventual discarding of modules, components, or entire products deemed impossible or excessively costly to recycle (Table 14.5).

The typical washing machine contains materials such that perhaps 60 percent by weight is immediately reusable or recyclable with no design changes. Examples include the steel frames and the mechanical components. The recyclability of much of the system is a chance occurrence, however, because insufficient thought has customarily been given at any design stage to the eventual potential or efficiency of recycling. Examples of this understandable omission are readily demonstrable:

- Nearly every subassembly, though not necessarily the side panels, is attached by using screws in threaded holes. This technique is time-consuming during both assembly and disassembly. Other industries, particularly the automotive industry, have shown that a variety of new pop-in/pop-out fasteners, quick-disconnect designs, and the like are quite secure and rigid while being much more efficient in installation and disassembly.
- There is significant nonuniformity in the plastics used in the washing machine, an approach that makes recycling much more difficult. Thermosets and thermoplastics of different kinds appear in various components manufactured by various suppliers, with no overall attempt at minimization of plastics diversity or recy-

TABLE 14.4 Customer Use

Element Designation	Element Value and Explanation
Matls. choice (4,1)	4 (No concerns.)
Energy use (4,2)	2 (Little energy minimization.)
Solid res.(4,3)	2 (No plan for defective part reuse.)
Liquid res.(4,4)	2 (Water use, detergent discharge.)
Gas res.(4,5)	4 (No concerns.)

TABLE 14.5 Refurbishment, Recycling, Disposal

Element Designation	Element Value and Explanation
Matls. choice (5,1)	2 (Lead presents toxicity complications.)
Energy use(5,2)	2 (No minimization of energy use at recycling.)
Solid res.(5,3)	1 (High materials diversity, little modularity, no ISO markings.)
Liquid res(5,4)	4 (No concerns.)
Gas res.(5,5)	3 (Some air pollution likely from metal recovery.)

cling efficiency. Regardless of the level of diversity, each plastic part should carry identification marking as designated by the International Standards Organization, but this is not common practice.

- Mixed materials often are joined in ways that are difficult or time-consuming to reverse on a recycling line. For example, enameled steel is a very difficult construction from which to recover all the constituent materials.
- No reuse plan is in place for electronic components.

14.3.2 SLCA Assessment Results

For convenience in summarizing the environmentally related characteristics of the generic washing machine, Table 14.6 collects the matrix assessment values for each of the five life-cycle stages and for the unit as a whole. The result is a score of 60 out of a possible 100 points. A comprehensive LCA would probably weight the product use phase heavily to reflect the long time span during which its environmental impacts occur, but the same design deficiencies would likely be singled out for attention by either the LCA or SLCA approach.

14.4 NEEDS-BASED APPROACHES

The preceding assessment of the typical washing machine has identified many areas in which its environmental attributes are less than ideal, chief among them being the use and discharge of energy, water, and detergents during use, the manufacture and subsequent need for recycling of packaging material by component suppliers and washing machine manufacturers, and the difficulties of materials recovery and recycling at end of product life. An alternative way to make the assessment is to perform the SLCA in

TABLE 14.6 The Environmentally Responsible Product Matrix for the Generic Washing Machine

Life Stage	Environmental Stressor					
	Materials Choice	Energy Use	Solid Residues	Liquid Residues	Gaseous Residues	Total
Pre-manufacture	0	2	2	3	2	9/20
Product manufacture	2	2	3	3	3	13/20
Product delivery	1	2	1	4	2	10/20
Product use	4	2	2	2	4	14/20
Refurbishment, recycling, disposal	2	2	1	4	3	12/20
Total	9/20	10/20	9/20	16/20	14/20	58/100

TABLE 14.7 The Environmentally Responsible Product Matrix for the Ideal Green Procurement of Clean Clothing

Life Stage	Environmental Stressor					
	Materials Choice	Energy Use	Solid Residues	Liquid Residues	Gaseous Residues	Total
Premanufacture	4	4	4	4	4	20/20
Product manufacture	4	4	4	4	4	20/20
Product delivery	4	4	4	4	4	20/20
Product use	4	4	4	4	4	20/20
Refurbishment, recycling, disposal	4	4	4	4	4	20/20
Total	20/20	20/20	20/20	20/20	20/20	100/100

reverse: the need rather than the product is the starting point, and desirable attributes of a product meeting that need are identified. From the standpoint of the environment, the ideal product would have a matrix assessment rating as shown in Table 14.7: perfect scores for all matrix elements. Given the traditional approach to the provisioning of clean clothing exemplified by the washing machine, is it possible to anticipate that a set of perfect scores can ever be achieved by this approach, while simultaneously providing customer satisfaction?

To begin with, it is unrealistic to imagine that any product can be manufactured while having absolutely no effect on the environment. The very nature of a product is that its manufacture involves the use of materials (thus involving resources) and generally their transport and transformation (thus involving energy). An ideal green product is therefore not one that has no environmental impact, but one that satisfies the customer need while meeting a long-term sustainability goal of having the absolute minimum environmental impact. Given this perspective, consider the premanufacture life stage. Matrix element 1,1 could be given a perfect score if it were possible to eliminate all toxic materials in supplier products. Because at present there are a variety of anti-corrosion coatings and conductive adhesives available, it is reasonable to anticipate eventual toxic material elimination. Similarly, we can anticipate the eventual recycling or elimination of supplier packaging (1,3), although the specific ways in which this may be accomplished are unclear, as is the development of delivery methods and transportation technologies that would greatly minimize transport-related impacts (1,2 and 1,5). Thus, in the premanufacture life stage there are significant technical challenges but no intractable constraints that would ultimately prevent the washing machine from becoming an ideal green product.

A similar conclusion is reached concerning the manufacturing life stage. The substitution for enamel of an alternative protective coating such as those used on automobile exteriors has the potential to essentially eliminate manufacturing energy losses (2,2), as well as atmospheric emissions (2,5) and related toxic concerns (2,1). Again, there appears to be nothing that cannot be improved in some way.

The product delivery stage is amenable to a number of fairly straightforward solutions to environmental concerns, at least in principle. One can imagine the elimination of packaging diversity (3,1), the recovery and reuse of packaging materials (3,3), and optimization of deliveries to point-of-sale and customer locations to reduce transport-related impacts to low levels (3,2 and 3,5). As before, these changes appear to be technologically feasible.

At the customer use life stage, new approaches to defective parts provisioning (longer mean lifetimes, efficient reuse of returned components) can probably minimize solid waste concerns (4,3). However, as long as clothes are washed in water using detergent and then dried, one can reduce but not eliminate the substantial use and depletion of energy and water (4,2 and 4,4) that is involved. These two requirements will forever make it difficult or impossible to design a perfectly green washing machine using traditional approaches.

Finally, look at end of product life. The toxicity concerns (5,1) can be eliminated given good materials choices by suppliers, and thoughtful design for disassembly is likely to minimize but not eliminate solid residues (5,3) and to reduce emissions (5,5) and energy use (5,2) to negligible levels. Hence, the environmental impacts of life stage 5 are open to improvement, but perhaps not to impact-free transformation.

The reverse life-cycle assessment has thus suggested that the traditional washing machine can eventually be greatly improved from an environmental standpoint, but that its environmental impacts can never be completely eliminated. This conclusion is reinforced when one takes account of the fact that most washing machines are used in conjunction with electric or gas dryers, appliances that also utilize resources and produce waste products. A similar result has been derived by looking at the problem from another standpoint: that of the environmental impacts attributable to an article of clothing. In a study performed by Franklin Associates for the American Fiber Manufacturers Association, energy use for laundering a woman's polyester blouse throughout its lifetime was estimated at some four times that used in manufacturing the garment, and solid residues related to municipal sludge and ash from off-site energy production were largely attributable to the consumer life stage as well. Improvements suggested were the use of cold wash cycles and line drying, both of which seem likely to result in decreased customer satisfaction.

From this discussion, it is apparent that in the typical washing machine one has a product created in response to a perceived need: The provisioning of clean clothing with maximum convenience and at affordable cost. From earliest civilization this need has been recognized, and it has been met throughout human history in essentially the same way—by treating garments as precious items to be preserved and maintained, and by using water and detergents to preserve and maintain them (mending and other repairs excepted). This system, already effective, could certainly be improved while taking the same basic approach to the need. There is no obvious reason why the desire for clean clothing can only be satisfied in this way, however. A "brainstorming" approach might imagine clothing that could be cleaned in other ways, as is the case currently for those fabrics that must be dry cleaned professionally. Alternatively, one could imagine manufacturing recyclable clothing so efficiently and inexpensively that clothing might never need to be cleaned at all, but could be recycled after wearing. It is worth examining the attributes of each of the three approaches.

Option 1 is the traditional approach, in which clothes are washed periodically in (generally heated) water, detergents, and perhaps other laundry products. A warm air dryer is not required as an adjunct to the washing machine, but in practice, at least in North America and much of Northern Europe, a dryer is almost always present. The latter requires electricity to operate, and uses either electricity or natural gas to heat the air for drying.

Certain fabrics perform better than others under numerous cycles of wearing, washing, and drying, and fabric manufacturers have evolved their products to be compatible with traditional washers and dryers. For some fabrics, however, suitable home systems have never been developed, and professional dry cleaning is required. Thus, taken all in all, a washing machine can be pictured as a component in a system that includes fiber growing and petroleum recovery, clothing manufacture, transport, sale, customer use, repeated washing and drying (or repeated dry cleaning), and disposal. It is the system is what one wants to optimize from an environmental standpoint, not just the washing machine.

The traditional clothes provisioning system has several advantages. The most obvious is that it is a system already in place. Consumers have purchased clothing and washers and dryers that together provide what they want: good-looking, relatively long-lasting, easily washed garments. For some fabrics, dry cleaning is needed and this infrastructure is in place as well. The disadvantages of the system are equally obvious: substantial use of energy and water occurs, chlorinated dry-cleaning solvents with potentially significant impacts on the workplace and the environment are emitted, and the technological approach is such that while these consequences may be mitigated, they can probably never be eliminated.

The second option is to retain the same basic approach we now take toward providing clean clothing, but to re-engineer that approach for optimum environmental performance. One can envision energy, water, and detergent use being markedly reduced. It may also be possible to optimize the rinse cycle and incorporate heat or radiation drying within the washer so as to eliminate the dryer altogether. This approach, following the "less is better" philosophy, would decrease, but not eliminate, the environmental effects of providing clean clothing.

Option 3 is a system in which clothing is cleaned without the use of water, but rather with some alternative technique such as infrared or microwave radiation. Liquid detergents might not be required; one could envision a system in which a detergent powder adheres to soil and serves as a locator for the radiant energy, for example. Because no wetting of clothing would be involved, no drying would be required. Dry cleaning for some fabrics might or might not still be needed.

Option 3 has as its principal advantages the elimination of the use of water, much of the energy, and perhaps detergents characteristic of the present system. Its principal disadvantage, of course, is that it has not been developed. A likely possibility is that such a system might be realized but only for specialized fabrics, in which case the clothing industry and its suppliers would have to be collaborators in the development of a new integrated system. Were the system to be realized technically, it would also be necessary to deploy it at a reasonable cost. Additionally, perhaps the most difficult requirement for any departure from common practice, it would have to be demonstrated that the new technique and the new clothing materials (if any) were at least as satisfactory to consumers as the old.

The fourth option that might be considered is the design and manufacture of "one-time" clothing, as we now do with other products such as adhesive bandages. The concept here is that clothing that meets customer needs would be manufactured, delivered, and sold in an efficient manner, worn once, and then recovered much as newspapers are routinely collected today. The recovered clothing would be reprocessed into newly-manufactured "one-time" clothing.

As with Option 3, the advantages of Option 4 are the elimination of energy and water use during washing and drying. The disadvantages are, first, that the Option 4 system has not been realized technically, and second, were it to be realized it would have to be satisfactory to customers. Option 4 would also require a drastic change in perspective that Option 3 would not: A switch to thinking of clothing not as a long-term investment but as a commodity to be purchased regularly, as is food. Indeed, many business people practice something close to this approach, because they take suits, shirts, blouses, and other garments to the dry cleaners once or twice a week. For them, at least, it would not be a great leap to stop at the clothing store rather than the dry cleaners. An additional advantage is that wardrobe changes would be quick and convenient—one could change clothing styles every week as easy as not.

14.5 DISCUSSION

This brief study has not encompassed several items that a complete study would treat. For example, our present washers and dryers were designed for cotton fabrics and blends, yet cotton growing is known to require large amounts of water and pesticides, and thus to be a heavy load on the environment. Polyester and other fibers are substantially better on that score and might fit neatly into an alternative cleaning or one-time system.

RLCA is not a tool that necessarily gives a ready answer to the question of which option is best from an environmental standpoint. In the example discussed here, Option 1 involves significant energy and water use and disposal. Option 2 is similar, but utilizes or produces less of each. Option 3 involves a new cleaning technology with uncertain physical and financial properties, and Option 4 trades off energy and water use on customer premises for additional manufacturing and product transport, with their attendant environmental impacts. In addition, a societal transformation is required for Option 4, and such transformations are typically difficult to achieve. Perhaps equally significant is the realization that time and investment are required for technological change, particularly those changes implied by Options 3 and 4, so the anticipated technological transition period must be part of the decision perspective.

However, the present purpose is not necessarily to select the best alternative for the provisioning of clean clothing, but to study the utility of the RLCA itself. Two interesting characteristics arise naturally out of performing a reverse life-cycle assessment. The first is that RLCA leads the designer into moving her or his thinking out of its traditional paths and into potentially more innovative ones. In all likelihood, some of these paths will be feasible and some not, a finding that may inspire research efforts to achieve feasibility for an approach that looks promising but is currently intractable. The second characteristic is that RLCA leads the designer into systems thinking, of trying to assess what approach to a societal need is most apposite from an overall perspective, rather than from the perspective of a small niche. RLCA is thus particularly appropriate for long-range corporate or governmental planning.

The above discussion has focused on cleaning clothing, but the discussion could readily be broadened to look at any generic cleaning process. In the early 1990s, for example, the use of CFCs for a variety of industrial cleaning operations came under scrutiny because of ozone layer concerns. The initial response of industry was to be more careful in the operation of existing processes—using smaller amounts, lowering cleaning temperatures to minimize CFC loss through volatilization, purifying and recycling CFCs, and the like—that is analogous to Option 2 for the cleaning of clothing. The next step was the development of alternative solvents, including limonene and *n*-butyl butyrate; not unlike Option 3 above. Finally, industry did more than *meet* the need for cleaning, it *evaluated* the need. In at least some applications, such as the cleaning of solder flux from electronic printed circuit boards, it was found that minor process changes could permit the cleaning step to be omitted entirely, an action along the lines of option 4.

Because of the technological and societal complexities of alternative ways of meeting a need, the best choice for meeting that need is not always obvious, and detailed study and consensus-building will generally be required. The real benefit of RLCA is not necessarily that it readily produces the right choice among options, but that it forces broad, innovative, systems thinking rather than attempting to achieve incremental improvements on an existing system without exploration of additional possibilities. The RLCA will not always be a tool that provides great new insights, but it has that potential, and should thus be a part of the arsenal of every industrial ecologist.

FURTHER READING

American Fiber Manufacturers Association, *Resource and Environmental Profile Analysis of a Manufactured Apparel Product: Woman's Knit Polyester Blouse*, Washington, D.C., 1993.

Dixcee, A., Washing machine design: The environmental dimension, *Industry and Environment* (Nairobi: United Nations Environment Programme, Oct.-Dec. 1993) pp. 43–45.

Environmental Protection Agency Science Advisory Board, *Beyond the Horizon: Using Foresight to Protect the Environmental Future*, Report SAB-EC-95-007 (Washington, D.C., 1995).

Ingle, K.A., *Reverse Engineering* (New York: McGraw-Hill, 1994).

Wheeler, D., What future for product lifecycle assessment? *Integrated Environmental Management*, 20 (1993) 15–19.

EXERCISES

14.1. Reverse LCAs start with the need served by a product rather than with the product itself. Identify the needs served by the following: (a) automobile; (b) filing cabinet; (c) lawn mower.

14.2. What are possible alternative ways of filling the needs identified in Exercise 14.1?

14.3. Apply reverse LCA thinking to the water supply and wastewater treatment infrastructure of a city. What needs are being served? What might be alternative ways of filling the needs?

CHAPTER 15

The Ultimate Life-Cycle Assessment

15.1 BREADTH OF ASSESSMENT

Today's life-cycle assessments suffer from a lack of breadth in their approach to environmental concerns. This lack of breadth arises from the attempts of LCA practitioners to generate detailed quantitative results as the primary assessment product. As a consequence, the environmental concerns that are treated in the assessments tend to be those that are readily reduced to numbers, such as the impact on the ozone layer of emitting a known quantity of CFCs. Those concerns that are inherently semiquantitative or wholly qualitative, such as a decrease in visibility due to photochemical smog or the loss of several species of songbirds, tend to be ignored or mentioned only in passing regardless of their importance.

A second driving force promoting quantification, i.e., the expression of the environment as a balance sheet, is the effort to internalize the costs of environmental degradation. The reason for this goal is to let normal marketplace functioning serve as protector of environmental resources and benefits. For the goal to be realized, however, a way must be found to evaluate the disappearance of a family of beetles or a tiny increase in the odds of suffering from emphysema. Because these evaluations are enormously contentious, little progress has been made in arriving at evaluations. Even if the task were to be accomplished, the world's economists would have to adopt the new approach to costing, a transition that would not occur easily.

Thus, while quantification where possible is certainly a good thing to do, it is important to recognize that qualification is an eminently practical and beneficial approach to environmental assessment. Once this perspective is accepted, one can proceed to expand the breadth of LCA/SLCAs to include all relevant environmental concerns, quantifiable or not. A reputable qualitative or semiquantitative assessment of the full spectrum of concerns is far preferable to a quantitative assessment of only a small fraction of them. The final product may not provide the definitive advice one would ideally like to have, but does provide information that can be used to compare alternatives and support decisions. From the environmental standpoint, it is far better to be approximately right today than exactly right a decade from now (should we indeed ever feel we are exactly right).

15.2 EFFICIENCY AND TRACTABILITY

This book has emphasized streamlined life-cycle assessments because their efficiency and tractability make it more probable that they will be employed on a regular basis.

That potential is enhanced if the assessment approach can be automated, and further enhanced if it can be embedded in the computer-aided design (CAD) tools used so commonly in design and development activities. While software tools have not been a topic of this book, a few comments on (S)LCA software are relevant.

The development of LCA/SLCA software is at early stages, but can nevertheless demonstrate some degree of usefulness. Perhaps the most significant is the calling out of environmentally related elements of design for special attention, regardless of whether or not the software provides judgments. Some programs automatically offer lists of alternative materials or components, an operation that is undeniably helpful. Others attempt to perform impact analyses of varying degrees of sophistication. Nonetheless, the implementation of completely satisfactory software remains a distant vision. This is not because the programming involved is particularly difficult; in fact, a number of programs have been developed, and some are commercially available. The difficulty is that such programs inevitably embed within themselves a number of contentious assumptions, most of which have been discussed in these pages. In a field as complex and evolving as LCA/SLCA, the tools must first be designed, then developed, and then honed to razor sharpness before they can be codified in a software package. Until that happens, and it is likely to be a long time, the expert assessor and not the computer programmer will play the crucial role in the LCA/SLCA process.

15.3 ALTERNATIVE TOOLS

When should one proceed with LCAs/SLCAs and when might other evaluation tools be more suitable? N.W. van den Berg and colleagues (see Further Reading) offer the following advice:

When to use LCAs/SLCAs:

- To compare the environmental impacts of different products with the same function.
- To compare the environmental impacts of one product with a reference or standard.
- To identify the environmentally most dominant stage in a product life cycle and hence to indicate the main routes to environmental improvements of existing products.
- To help in the design of new products and services.
- To indicate strategically the direction of development.

This list is reasonable, straightforward, and good to keep in mind. Also offered, however, is a second list, which reads (in part):

When Not to Use LCAs/SLCAs and What Else to Use:

- Not to answer questions related to a single production process. Instead use a process technology study.
- Not to answer specific environmental questions regarding geographical locations. Instead use environmental impact assessment.

- Not to answer questions related to the environmental impacts of a company. Instead use an environmental audit.

Unlike the first list, the second is much more problematic because of deficiencies in the tools that are suggested, primarily their restriction to limited life stages and selected environmental concerns. Environmental impact assessments, for example, rarely or never consider long temporal or spatial scales, or the final life stage. Environmental audits are normally confined to current activities under direct corporate control, with past and future impacts and upstream and downstream impacts being outside the scope of study. Similar concerns apply to process technology studies. Conversely, SLCA approaches are shown in previous chapters to be suitable for all the applications traditionally evaluated by these alternative tools. Thus, LCAs/SLCAs, if properly constructed, are applicable in *all* the instances cited above.

15.4 THE INDUSTRIAL ECOLOGIST AS PHYSICIAN

Life-cycle assessment has been heralded as the way to ensure environmentally responsible products, and advocates promote the concept that its results are of great utility once the needed data are at hand. While the benefits of evaluating products throughout their life cycles and over the spectrum of environmental concerns is undisputed, the integrity and completeness of the results has frequently been questioned on grounds that have been discussed earlier, especially in Chapter 5. Where does this leave the responsible technologist assigned to evaluate products, processes, and facilities? The answer is that life-cycle assessment should be approached from the standpoint of modern medicine as much as from that of modern technology, because assessment, like medicine, is more than a technological activity.

Review, for a moment, the medical approach to diagnosis. As with LCA, a significant amount of information is available, but a significant amount is not. The physician has at his or her disposal the results of certain tests, the patient's description of symptoms, and the physician's own observations of the patient. Additional information that might be useful—additional blood analyses, biopsy reports, CAT scans—may not be available because of patient condition, logistical difficulties, and the like. Still other potentially useful knowledge—identification of a deadly virus, specification of diagnostic tests for failure of specific immunosuppressive functions—may be impossible to acquire, at least in the time available. Nonetheless, physicians make diagnoses. They have been doing so since before they understood the circulation of the blood, before they characterized the functions of the immune system, and before DNA testing was available. Their diagnoses, on average, have undoubtedly improved with time, and no physician would claim them to be perfect or anticipate that they will ever be; the human body is simply too complex and the analytic probes too few for that. Despite these limitations, physicians diagnose, make recommendations, and most of the patients are the better for it. Physicians are probably more respected because of this need to apply judgment than they would be if they simply took their diagnoses from a computer.

The situation of the industrial ecologist confronting a life-cycle assessment is similar in a number of ways to that of a physician confronting a patient, and quite dissimilar to that of an auto mechanic confronting a misperforming engine. Industrial ecologists will never have all the data they need—some is difficult or impossible to get, some must be treated differently in different places and over different time scales, some is societally dependent, and some will be negated by new discoveries in the environmental sciences. Enough is available, however, for diagnoses to occur, particularly with the aid of expert systems approaches that have been so successful in medicine and elsewhere.

15.5 THE ULTIMATE LCA

The ultimate LCA is one that is approached in a reputable fashion by an expert practitioner, completed promptly and professionally, and whose recommendations are implemented. All three of these stages need to happen if the exercise is to have its maximum value. The scope and detail will differ depending on the customer, but each assessment might well proceed along the following lines:

- A reverse LCA should be conducted as the first step in the process. Careful consideration should be given to filling the perceived need in non-traditional ways, provided those ways are likely to be environmentally beneficial.
- After the approach to the need is determined, the LCA/SLCA is then conducted. In the vast majority of cases, the simplified form of the matrix LCA described in Chapters 7–10 will be adequate for providing guidance to the product or process designer, the architect, or the service provider. The minimal useful activity in this regard is to "follow the energy," as energy use is often a reasonable surrogate for overall environmental performance.
- For activities that are large, or for which designs will be replicated in identical or similar fashion, consideration should be given to performing an enhanced SLCA, as described in Chapter 13. The intellectual rigor will be improved over the simple approach, and the additional expenditure of time and money will be small because no additional data are required.
- For very large or very important activities, such as global resource budgeting or global assessments of industrial sector environmental interactions or for academic explorations of industrial processes, a formal, comprehensive LCA may be considered. Doing so is the most intellectually rigorous of the life-cycle approaches that are available; the penalty one pays is considerable additional expenditure of effort on data collection and analysis.
- To illustrate the concepts that are discussed in this book, a particular matrix tool has been employed in several different guises. As seen in Chapter 6, however, a number of similar SLCA tools are extant. Regardless of the precise tool used for assessment, the bottom line is whether or not the recommendations that result are implemented. It is at the implementation stage that the LCA/SLCA analyst becomes sales person, presenting the results succinctly, prioritizing the recom-

mendations, working closely with the decision makers, and having a good feel for which changes can be made easily and inexpensively and which will be costly and difficult. This stage, often not emphasized, is the key to the utility of the LCA/SLCA process.

It is appropriate to conclude this book by conceding that LCA/SLCA is not strictly a rigorous activity, and that therefore analysts are not cogs in a well-understood wheel, but rather might ultimately be termed planetary physicians. Their advice will, of necessity, be given often in the face of a scarcity of information: industrial, environment, and social. Some of their diagnoses will be straightforward, others will not. New planetary diseases will emerge to challenge them, and new solutions for the diseases will be proposed. The path will not be straight. Nonetheless, most of their diagnoses will do some good for the ultimate patient—the planet on which we all live. It is difficult to imagine a more useful role, or a higher calling.

FURTHER READING

Schmidheiny, S., *Changing Course: A Global Business Perspective on Development and the Environment* (Cambridge, MA, The MIT Press, 1992).

Smart, B., Ed., *Beyond Compliance: A New Industry View of the Environment* (Washington, D.C.: World Resources Institute, 1992).

Van den Berg, N.W., G. Huppes, and C.E. Dutilh, Beginning LCA: A Dutch guide to environmental life-cycle assessment, in *Environmental Life-Cycle Assessment*, M.A. Curran, Ed. (New York: McGraw-Hill, 1996) pp. 17.1–17.41.

APPENDIX A

Environmentally Responsible Product Matrix: Scoring Guidelines and Protocols

The Product Improvement Matrix is described in Chapter 7. In this appendix, a sample of possible scoring considerations appropriate to each of the matrix elements is presented. It is anticipated that different types of products may require different checklists and evaluations, so this appendix is presented as an example rather than as a universal formula.

PRODUCT MATRIX ELEMENT: 1,1

Life Stage: Premanufacture
Environmental Stressor: Materials Choice

If any the following conditions apply, the matrix element rating is 0:

- For the case where supplier components/subsystems are used: No/little information is known about the chemical content in supplied products and components.
- For the case where materials are acquired from suppliers: A scarce material is used where a reasonable alternative is available. (Scarce materials are defined as antimony, beryllium, boron, cobalt, chromium, gold, mercury, the platinum metals [Pt, Ir, Os, Pa, Rh, Ru], silver, thorium, and uranium.)

If the following condition applies, the matrix element rating is 4:

- No virgin material is used in incoming components or materials.

If neither of the preceding ratings is assigned, complete the checklist below. Assign a rating of 1, 2, or 3 depending on the degree to which the product meets DFE preferences for this matrix element.

- Is the product designed to minimize the use of materials in restricted supply (see preceding list)?
- Is the product designed to utilize recycled materials or components wherever possible?

PRODUCT MATRIX ELEMENT: 1,2

Life Stage: Premanufacture
Environmental Stressor: Energy Use

If the following condition applies, the matrix element rating is 0:

- One or more of the principal materials used in the product requires energy-intensive extraction and suitable alternative materials are available that do not. (Materials requiring energy-intensive extraction are defined as virgin aluminum, virgin steel, and virgin petroleum.)

If the following condition applies, the matrix element rating is 4:

- Negligible energy is needed to extract or ship the materials or components for this product.

If neither of the preceding ratings is assigned, complete the checklist below. Assign a rating of 1, 2, or 3 depending on the degree to which the product meets DFE preferences for this matrix element.

- Is the product designed to minimize the use of virgin materials whose extraction is energy-intensive?
- Does the product design avoid or minimize the use of high-density materials whose transport to and from the facility will require significant energy use? (Such materials are defined as those with a specific gravity preceding 7.0.)
- Is transport distance of incoming materials and components minimized?

PRODUCT MATRIX ELEMENT: 1,3

Life Stage: Premanufacture
Environmental Stressor: Solid Residues

If either of the following conditions apply, the matrix element rating is 0:

- For the case where materials are acquired from suppliers: Metals from virgin ores are used, creating substantial waste rock residues that could be avoided by the use of recycled material, and suitable recycled material is available from recycling streams.
- For the case where supplier components/subsystems are used: All incoming packaging is from virgin sources and consists of three or more types of materials.

If all of the following conditions apply, the matrix element rating is 4:

- For the case where materials are acquired from suppliers: No solid residues result from resource extraction or during production of materials by recycling (example: petroleum).
- For the case where supplier components/subsystems are used: None/minimal packaging material is used or supplier takes back all packaging material.

Appendix A   Environmentally Responsible Product Matrix   237

- For the case where supplier components/subsystems are used: Incoming packaging is totally reused/recycled.

If neither of the preceding ratings is assigned, complete the checklist below. Assign a rating of 1, 2, or 3 depending on the degree to which the product meets DFE preferences for this matrix element.

- Is the product designed to minimize the use of materials whose extraction or purification involves the production of large amounts of solid residues (i.e., coal and all virgin metals)?
- Is the product designed to minimize the use of materials whose extraction or purification involves the production of toxic solid residues? (This category includes all radioactive materials.)
- Has incoming packaging volume and weight, at and among all levels (primary, secondary, and tertiary), been minimized?
- Is materials diversity minimized in incoming packaging?

PRODUCT MATRIX ELEMENT: 1,4

Life Stage: Premanufacture
Environmental Stressor: Liquid Residues

If either of the following conditions apply, the matrix element rating is 0:

- For the case where supplier components/subsystems are used: Metals from virgin ores that cause substantial acid mine drainage are used, and suitable virgin material is available from recycling streams. (Materials causing acid mine drainage are defined as copper, iron, nickel, lead, and zinc.)
- For the case where materials are acquired from suppliers: The packaging contains toxic or hazardous substances that might leak from it if improper disposal occurs.

If both of the following conditions apply, the matrix element rating is 4:

- For the case where materials are acquired from suppliers: No liquid residues result from resource extraction or during production of materials by recycling.
- For the case where supplier components/subsystems are used: No liquid residue is generated during transportation, unpacking, or use of this product.

If neither of the preceding ratings is assigned, complete the checklist below. Assign a rating of 1, 2, or 3 depending on the degree to which the product meets DFE preferences for this matrix element.

- Is the product designed to minimize the use of materials whose extraction or purification involves the generation of large amounts of liquid residues? (This category includes paper and allied products, coal, and materials from biomass.)
- Is the product designed to minimize the use of materials whose extraction or purification involves the generation of toxic liquid residues? (These materials are defined as aluminum, chronimum, copper, iron, lead, mercury, nickel, and zinc.)

- Are refillable/reusable containers used for incoming liquid materials where appropriate?
- Does the use of incoming components require cleaning that involves a large amount of water or that generate liquid residues needing special disposal methods?

PRODUCT MATRIX ELEMENT: 1,5

Life Stage: Premanufacture
Environmental Stressor: Gaseous Residues

If the following condition applies, the matrix element rating is 0:

- The materials used cause substantial emissions of toxic, smog-producing, or greenhouse gases into the environment, and suitable alternatives that do not do so are available. (These materials are defined as aluminum, chromium, copper, iron, lead, mercury, nickel, zinc, paper and allied products, and concrete.)

If the following condition applies, the matrix element rating is 4:

- No gaseous residues are produced during resource extraction or production of materials by recycling.

If neither of the preceding ratings is assigned, complete the checklist below. Assign a rating of 1, 2, or 3 depending on the degree to which the product meets DFE preferences for this matrix element.

- Is the product designed to minimize the use of materials whose extraction or purification involves the generation of large amounts of gaseous (toxic or otherwise) residues? (Such materials are defined as aluminum, copper, iron, lead, nickel, and zinc.)

PRODUCT MATRIX ELEMENT: 2,1

Life Stage: Product Manufacture
Environmental Stressor: Materials Choice

If the following condition applies, the matrix element rating is 0:

- Product manufacture requires relatively large amounts of materials that are restricted (see [1,1]), toxic, and/or radioactive.

If the following condition applies, the matrix element rating is 4:

- Materials used in manufacture are completely closed loop (captured and reused/recycled) with minimum inputs required.

If neither of the preceding ratings is assigned, complete the checklist below. Assign a rating of 1, 2, or 3 depending on the degree to which the product meets DFE preferences for this matrix element.

- Do manufacturing processes avoid the use of materials that are in restricted supply?
- Is the use of toxic material avoided or minimized?
- Is the use of radioactive material avoided?
- Is the use of virgin material minimized?
- Has the chemical treatment of materials and components been minimized

PRODUCT MATRIX ELEMENT: 2,2

Life Stage: Product Manufacture
Environmental Stressor: Energy Use

If the following condition applies, the matrix element rating is 0:

- Energy use for product manufacture/testing is high and less energy-intensive alternatives are available.

If the following condition applies, the matrix element rating is 4:

- Product manufacture and testing requires no or minimal energy use.

If neither of the preceding ratings is assigned, complete the checklist below. Assign a rating of 1, 2, or 3 depending on the degree to which the product meets DFE preferences for this matrix element.

- Is the product manufacture designed to minimize the use of energy-intensive processing steps?
- Is the product manufacture designed to minimize energy-intensive evaluation/testing steps?
- Do the manufacturing processes use co-generation, heat exchanges, and/or other techniques to utilize otherwise waste energy?
- Is the manufacturing facility powered down when not in use?

PRODUCT MATRIX ELEMENT: 2,3

Life Stage: Product Manufacture
Environmental Stressor: Solid Residues

If the following condition applies, the matrix element rating is 0:

- Solid manufacturing residues are large and no reuse/recycling programs are in use.

If the following condition applies, the matrix element rating is 4:

- Solid manufacturing residues are minor and each constituent is >90 percent reused/recycled.

If neither of the preceding ratings is assigned, complete the checklist below. Assign a rating of 1, 2, or 3 depending on the degree to which the product meets DFE preferences for this matrix element.

- Have solid manufacturing residues been minimized and reused to the greatest extent possible?
- Has the resale of all solid residues as inputs to other products/processes been investigated and implemented?
- Are solid manufacturing residues that do not have resale value minimized and recycled?

PRODUCT MATRIX ELEMENT: 2,4

Life Stage: Product Manufacture
Environmental Stressor: Liquid Residues

If the following condition applies, the matrix element rating is 0:

- Liquid manufacturing residues are large and no reuse/recycling programs are in use.

If the following condition applies, the matrix element rating is 4:

- Liquid manufacturing residues are minor and each constituent is >90 percent reused/recycled.

If neither of the preceding ratings is assigned, complete the checklist below. Assign a rating of 1, 2, or 3 depending on the degree to which the product meets DFE preferences for this matrix element.

- If solvents or oils are used in the manufacture of this product, is their use minimized and have alternatives been investigated and implemented?
- Have opportunities for sale of all liquid residues as input to other processes/products been investigated and implemented?
- Have the processes been designed to require the maximum recycled liquid process chemicals rather than virgin materials?

PRODUCT MATRIX ELEMENT: 2,5

Life Stage: Product Manufacture
Environmental Stressor: Gaseous Residues

If either of the following conditions apply, the matrix element rating is 0:

- Gaseous manufacturing residues are large and no reuse/recycling programs are in use.
- CFCs are used in product manufacture.

If the following condition applies, the matrix element rating is 4:

- Gaseous manufacturing residues are relatively minor and reuse/recycling programs are in use.

Appendix A Environmentally Responsible Product Matrix   241

If neither of the preceding ratings is assigned, complete the checklist below. Assign a rating of 1, 2, or 3 depending on the degree to which the product meets DFE preferences for this matrix element.

- If HCFCs are used in the manufacture of this product, have alternatives been thoroughly investigated and implemented?
- Are greenhouse gases used or generated in any manufacturing process connected with this product?
- Have the resale of all gaseous residues as inputs to other processes/products been investigated and implemented?

PRODUCT MATRIX ELEMENT: 3,1

Life Stage: Product Packaging and Transport
Environmental Stressor: Materials Choice

If the following condition applies, the matrix element rating is 0:

- All outgoing packaging is from virgin sources and consists of three or more types of materials.

If the following condition applies, the matrix element rating is 4:

- No outgoing packaging or minimal and recycled packaging material is used.

If neither of the preceding ratings is assigned, complete the checklist below. Assign a rating of 1, 2, or 3 depending on the degree to which the product meets DFE preferences for this matrix element.

- Does the product packaging minimize the number of different materials used and is it optimized for weight/volume efficiency?
- Have efforts been made to use recycled materials for product packaging and to make sure the resulting package is recyclable and marked as such?
- Is there a functioning recycling infrastructure for the product packaging material?
- Have the packaging engineer and the installation personnel been consulted during the product design?

PRODUCT MATRIX ELEMENT: 3,2

Life Stage: Product Packaging and Transportation
Environmental Stressor: Energy Use

If the following condition applies, the matrix element rating is 0:

- Packaging material extraction, packaging procedure, and transportation/installation method(s) are all energy intensive and less energy-intensive options are available.

If the following condition applies, the matrix element rating is 4:

- Packaging material extraction, packaging procedure, and transportation/installation methods(s) all require little or no energy.

If neither of the preceding ratings is assigned, complete the checklist below. Assign a rating of 1, 2, or 3 depending on the degree to which the product meets DFE preferences for this matrix element.

- Do packaging procedures avoid energy-intensive activities?
- Are component supply systems and product distribution/installation plans designed to minimize energy use?
- If installation is involved, is it designed to avoid energy-intensive procedures?
- Is long distance, energy-intensive product transportation avoided or minimized?

PRODUCT MATRIX ELEMENT: 3,3

Life Stage: Product Packaging and Transportation
Environmental Stressor: Solid Residues

If the following condition applies, the matrix element rating is 0:

- Outgoing packaging material is excessive, with little consideration given to recycling or reuse.

If the following condition applies, the matrix element rating is 4:

- Minimal or no outgoing packaging material is used and/or the packaging is totally reused or recycled.

If neither of the preceding ratings is assigned, complete the checklist below. Assign a rating of 1, 2, or 3 depending on the degree to which the product meets DFE preferences for this matrix element.

- Is the product packaging designed to make it easy to separate the constituent materials?
- Do the packaging materials need special disposal after products are unpacked?
- Has product packaging volume and weight, at and among all three levels, (primary, secondary, and tertiary) been minimized?
- Are arrangements made to take back product packaging for reuse and/or recycling?
- Is materials diversity minimized in outgoing product packaging?

PRODUCT MATRIX ELEMENT: 3,4

Life Stage: Product Packaging and Transportation
Environmental Stressor: Liquid Residues

If the following condition applies, the matrix element rating is 0:

- The product packaging contains toxic or hazardous substance that might leak from it if improper disposal occurs (such as the acid from batteries).

If the following condition applies, the matrix element rating is 4:

- Little or no liquid residue is generated during packaging, transportation, or installation of this product.

If neither of the preceding ratings is assigned, complete the checklist below. Assign a rating of 1, 2, or 3 depending on the degree to which the product meets DFE preferences for this matrix element.

- Are refillable or reusable containers used for liquid products where appropriate?
- Do the product packaging operations need cleaning/maintenance procedures that require a large amount of water or generate other liquid residues (oils, detergents, etc.) that need special methods for disposal?
- Do the product unpacking and/or installation operations require cleaning that involves a large amount of water or that generate liquid residues needing special disposal methods?

PRODUCT MATRIX ELEMENT: 3,5

Life Stage: Product Packaging and Transportation
Environmental Stressor: Gaseous Residues

If the following condition applies, the matrix element rating is 0:

- Abundant gaseous residues are generated during packaging, transportation, or installation, and alternative methods that would significantly reduce gaseous emissions are available.

If the following condition applies, the matrix element rating is 4:

- Little or no gaseous residues are generated during packaging, transportation, or installation of this product.

If neither of the preceding ratings is assigned, complete the checklist below. Assign a rating of 1, 2, or 3 depending on the degree to which the product meets DFE preferences for this matrix element.

- If the product contains pressurized gases, are transport/installation procedures designed to avoid their release?
- Are product distribution plans designed to minimize gaseous emissions from transport vehicles?
- If the packaging is recycled for its energy content (i.e., incinerated), have the materials been selected to ensure no toxic gases are released?

PRODUCT MATRIX ELEMENT: 4,1

Life Stage: Product Use
Environmental Stressor: Materials Choice

If the following condition applies, the matrix element rating is 0:

- Consumables contain significant quantities of materials in restricted supply or toxic/hazardous substances.

If the following condition applies, the matrix element rating is 4:

- Product use and product maintenance requires no consumables.

If neither of the preceding ratings is assigned, complete the checklist below. Assign a rating of 1, 2, or 3 depending on the degree to which the product meets DFE preferences for this matrix element.

- Has consumable material use been minimized?
- If the product is designed to be disposed of after using, have alternative approaches for accomplishing the same purpose been examined?
- Have the materials been chosen such that no environmentally inappropriate maintenance is required, and no unintentional release of toxic materials to the environment occurs during use?
- Are consumable materials generated from recycled streams rather than virgin material?

PRODUCT MATRIX ELEMENT: 4,2

Life Stage: Product Use
Environmental Stressor: Energy Use

If the following condition applies, the matrix element rating is 0:

- Product use and/or maintenance is relatively energy intensive and less energy-intensive methods are available to accomplish the same purpose.

If the following condition applies, the matrix element rating is 4:

- Product use and maintenance requires little or no energy.

If neither of the preceding ratings is assigned, complete the checklist below. Assign a rating of 1, 2, or 3 depending on the degree to which the product meets DFE preferences for this matrix element.

- Has the product been designed to minimize energy use while in service?
- Has energy use during maintenance/repair been minimized?
- Have energy-conserving design features (such as auto shutoff or enhanced insulation) been incorporated?
- Can the product monitor and display its energy use and/or its operating energy efficiency while in service?

PRODUCT MATRIX ELEMENT: 4,3

Life Stage: Product Use
Environmental Stressor: Solid Residues

If the following condition applies, the matrix element rating is 0:

- Product generates significant quantities of hazardous/toxic solid residue during use or from repair/maintenance operations.

If the following condition applies, the matrix element rating is 4:

- Product generates no (or relative minor amounts of) solid residue during use or from repair/maintenance operations.

If neither of the preceding ratings is assigned, complete the checklist below. Assign a rating of 1, 2, or 3 depending on the degree to which the product meets DFE preferences for this matrix element.

- Has the periodic disposal of solid materials (such as cartridges, containers, or batteries) associated with the use and/or maintenance of this product been avoided or minimized?
- Have alternatives to the use of solid consumables been thoroughly investigated and implemented where appropriate?
- If intentional dissipative emissions to land occur as a result of using this product, have less environmentally harmful alternatives been investigated?

PRODUCT MATRIX ELEMENT: 4,4

Life Stage: Product Use
Environmental Stressor: Liquid Residues

If the following condition applies, the matrix element rating is 0:

- Product generates significant quantities of hazardous/toxic liquid residue during use or from repair/maintenance operations.

If the following condition applies, the matrix element rating is 4:

- Product generates no (or relatively minor amounts of) liquid residues during use or from repair/maintenance operations.

If neither of the preceding ratings is assigned, complete the checklist below. Assign a rating of 1, 2, or 3 depending on the degree to which the product meets DFE preferences for this matrix element.

- Has the periodic disposal of liquid materials (such as lubricants and hydraulic fluids) associated with the use and/or maintenance of this product been avoided or minimized?
- Have alternatives to the use of liquid consumables been thoroughly investigated and implemented where appropriate?

- If intentional dissipative emissions to water occur as a result of using this product, have less environmentally harmful alternatives been investigated?
- If product contains liquid material that has the potential to be unintentionally dissipated during use or repair, have appropriate preventive measures been incorporated?

PRODUCT MATRIX ELEMENT: 4,5

Life Stage: Product Use
Environmental Stressor: Gaseous Residues

If the following condition applies, the matrix element rating is 0:

- Product generates significant quantities of hazardous/toxic gaseous residue during use or from repair/maintenance operations

If the following condition applies, the matrix element rating is 4:

- Product generates no (or relatively minor amounts of) gaseous residues during use or from repair/maintenance operations.

If neither of the preceding ratings is assigned, complete the checklist below. Assign a rating of 1, 2, or 3 depending on the degree to which the product meets DFE preferences for this matrix element.

- Has the periodic emission of gaseous materials (such as CO_2, SO_2, VOCs, and CFCs) associated with the use and/or maintenance of this product been avoided or minimized?
- Have alternatives to the use of gaseous consumables been thoroughly investigated and implemented where appropriate?
- If intentional dissipative emissions to air occur as a result of using this product, have less environmentally harmful alternatives been investigated?
- If product contains any gaseous materials that have the potential to be unintentionally dissipated during use or repair, have the appropriate preventive measures been incorporated?

PRODUCT MATRIX ELEMENT: 5,1

Life Stage: Recycling, Disposal
Environmental Stressor: Material Choice

If the following condition applies, the matrix element rating is 0:

- Product contains significant quantities of mercury (i.e., mercury relays), asbestos (i.e., asbestos based insulations), or cadmium (i.e., cadmium or zinc plated parts) that are not clearly identified and easily removable.

If the following condition applies, the matrix element rating is 4:

- Material diversity is minimized, the product is easy to disassemble, and all parts are recyclable.

If neither of the preceding ratings is assigned, complete the checklist below. Assign a rating of 1, 2, or 3 depending on the degree to which the product meets DFE preferences for this matrix element.

- Have materials been chosen and used in light of the desired recycling/disposal option for the product (e.g., for incineration, for recycling, for refurbishment)?
- Does the product minimize the number of different materials that are used in its manufacture?
- Are the different materials easy to identify and separate?
- Is this a battery-free product?
- Is this product free of components containing PCBs or PCTs (e.g., incapacitors and transformers)?
- Are major plastics parts free of polybrominated flame retardants or heavy metal-based additives (colorants, conductors, stabilizers, etc.)?

PRODUCT MATRIX ELEMENT: 5,2

Life Stage: Recycling, Disposal
Environmental Stressor: Energy Use

If the following condition applies, the matrix element rating is 0:

- Recycling/disposal of this product is relatively energy intensive (compared to other products that perform the same function) due to its weight, construction, and/or complexity.

If the following condition applies, the matrix element rating is 4:

- Energy use for recycling or disposal of this product is minimal.

If neither of the preceding ratings is assigned, complete the checklist below. Assign a rating of 1, 2, or 3 depending on the degree to which the product meets DFE preferences for this matrix element.

- Is the product designed with the aim of minimizing the use of energy-intensive process steps in disassembly?
- Is the product designed for high-level reuse of materials? (Direct reuse in a similar product is preferable to a degraded reuse.)
- Will transport of products for recycling be energy intensive because of product weight or volume or the location of recycling facilities?

PRODUCT MATRIX ELEMENT: 5,3

Life Stage: Recycling, Disposal
Environmental Stressor: Solid Residues

If the following condition applies, the matrix element rating is 0:

- Product consists primarily of unrecyclable solid materials (such as rubber, fiberglass, and compound polymers).

If the following condition applies, the matrix element rating is 4:

- Product can be easily refurbished and reused and is easily dismantled and 100 percent reused/recycled at the end of its life. For example, no part of this product will end up in a landfill.

If neither of the preceding ratings is assigned, complete the checklist below. Assign a rating of 1, 2, or 3 depending on the degree to which the product meets DFE preferences for this matrix element.

- Has the product been assembled with fasteners such as clips or hook-and-loop attachments rather than chemical bonds (gels, potting compounds) or welds?
- Have efforts been made to avoid joining dissimilar materials together in ways difficult to reverse?
- Are all plastic components identified by ISO markings as to their content?
- If product consists of plastic parts is there one dominant (>80 percent by weight) species?
- Is this product to be leased rather than sold?

PRODUCT MATRIX ELEMENT: 5,4

Life Stage: Recycling, Disposal
Environmental Stressor: Liquid Residues

If the following condition applies, the matrix element rating is 0:

- Product contains primarily unrecyclable liquid materials.

If the following condition applies, the matrix element rating is 4:

- Product uses no operating liquids (such as oils, coolants, or hydraulic fluids) and no cleaning agents or solvents are necessary for its reconditioning.

If neither of the preceding ratings is assigned, complete the checklist below. Assign a rating of 1, 2, or 3 depending on the degree to which the product meets DFE preferences for this matrix element.

- Can liquids contained in the product be recovered at disassembly rather than lost?
- Does disassembly, recovery and reuse generate liquid residues?
- Does materials recovery and reuse generate liquid residues?

PRODUCT MATRIX ELEMENT: 5,5

Life Stage: Recycling, Disposal
Environmental Stressor: Gaseous Residues

If the following condition applies, the matrix element rating is 0:

- Product contains or produces primarily unrecyclable gaseous materials that are dissipated to the atmosphere at the end of its life.

If the following condition applies, the matrix element rating is 4:

- Product contains no substances lost to evaporation/sublimation (other than water) and no volatile substances are used for refurbishment.

If neither of the preceding ratings is assigned, complete the checklist below. Assign a rating of 1, 2, or 3 depending on the degree to which the product meets DFE preferences for this matrix element.

- Can gases contained in the product be recovered at disassembly rather than lost?
- Does materials recovery and reuse generate gaseous residues?
- Can plastic parts be incinerated without requiring sophisticated air pollution control devices? Plastic parts that can cause difficulty in this regard are those that contain polybrominated flame retardants or metal based additives, are finished with polyurethane based paints, or are plated or painted with metals.

APPENDIX B: Environmentally Responsible Process Matrix: Scoring Guidelines and Protocols

The Process Improvement Matrix is described in Chapter 8. In this appendix, a sample of possible scoring considerations appropriate to each of the matrix elements is presented. It is anticipated that different types of processes may require different check lists and evaluations, so this appendix is presented as an example rather than as a universal formula.

PROCESS MATRIX ELEMENT: 1,1

Life Stage: Resource Provisioning
Environmental Stressor: Materials Choice

If either of the following conditions apply, the matrix element rating is 0:

- For the case where supplier components/subsystems are used: No/little information is known about the chemical content in supplied process consumables and equipment.
- For the case where materials are acquired from suppliers: A scarce material is used where a reasonable alternative is available. (Scarce materials are defined as antimony, beryllium, boron, cobalt, chromium, gold, mercury, the platinum metals [Pt, Ir, Os, Pa, Rh, Ru], silver, thorium, and uranium.)

If the following condition applies, the matrix element rating is 4:

- No virgin material is used in incoming components or materials.

If neither of the preceding ratings is assigned, complete the checklist below. Assign a rating of 1, 2, or 3 depending on the degree to which the process meets DFE preferences for this matrix element.

- Is the process designed to minimize the use of materials in restricted supply?
- Is the process designed to utilize recycled materials or components wherever possible?

Appendix B   Environmentally Responsible Process Matrix   251

- Of the potential consumable materials, are those chosen the ones whose extraction results in the lowest environmental impact?

PROCESS MATRIX ELEMENT: 1,2

Life Stage: Resource Provisioning
Environmental Stressor: Energy Use

If the following condition applies, the matrix element rating is 0:

- One or more of the principal materials used in the process requires energy-intensive extraction and suitable alternative materials are available that do not. (Materials requiring energy-intensive extraction are defined as virgin aluminum, virgin steel, and virgin petroleum.)

If the following condition applies, the matrix element rating is 4:

- Negligible energy is needed to extract or ship the materials or components for this process.

If neither of the preceding ratings is assigned, complete the checklist below. Assign a rating of 1, 2, or 3 depending on the degree to which the process meets DFE preferences for this matrix element.

- Is the process designed to minimize the use of virgin materials whose extraction is energy intensive?
- Does the process design avoid or minimize the use of high-density materials whose transport to and from the facility will require significant energy use? (Such materials are defined as those with a specific gravity above 7.0.)
- Is transport distance of incoming materials and components minimized?

PROCESS MATRIX ELEMENT: 1,3

Life Stage: Resource Provisioning
Environmental Stressor: Solid Residues

If either of the following conditions apply, the matrix element rating is 0:

- For the case where materials are acquired from suppliers: Metals from virgin ores are used, creating substantial waste rock residues that could be avoided by the use of virgin material, and suitable virgin material is available from recycling streams.
- For the case where supplier components/subsystems are used: All incoming packaging is from virgin sources and consists of three or more types of materials.

If all of the following conditions apply, the matrix element rating is 4:

- For the case where materials are acquired from suppliers: No solid residues result from resource extraction or during production of materials by recycling (example: petroleum).

- For the case where supplier components/subsystems are used: Minimal or no packaging material is used or the supplier takes back all packaging material.
- For the case where supplier components/subsystems are used: Incoming packaging is totally reused/recycled.

If neither of the preceding ratings is assigned, complete the checklist below. Assign a rating of 1, 2, or 3 depending on the degree to which the process meets DFE preferences for this matrix element.

- Is the process designed to minimize the use of materials whose extraction or purification involves the production of large amounts of solid residues (i.e., coal and all virgin metals)?
- Is the process designed to minimize the use of materials whose extraction or purification involves the production of toxic solid residues? (This category includes all radioactive materials.)
- Has incoming packaging volume and weight, at and among all levels (primary, secondary, and tertiary), been minimized?
- Is materials diversity minimized in incoming packaging?

PROCESS MATRIX ELEMENT: 1,4

Life Stage: Resource Provisioning
Environmental Stressor: Liquid Residues

If either of the following conditions apply, the matrix element rating is 0:

- For the case where supplier components/subsystems are used: Metals from virgin ores that cause substantial acid mine drainage are used, and suitable virgin material is available from recycling streams. (Materials causing acid mine drainage are defined as copper, iron, nickel, lead, and zinc.)
- For the case where materials are acquired from suppliers: The packaging contains toxic or hazardous substances that might leak from it if improper disposal occurs.

If both of the following conditions apply, the matrix element rating is 4:

- For the case where materials are acquired from suppliers: No liquid residues result from resource extraction or during production of materials by recycling.
- For the case where supplier components/subsystems are used: No liquid residue is generated during transportation, unpacking, or use of this product.

If neither of the preceding ratings is assigned, complete the checklist below. Assign a rating of 1, 2, or 3 depending on the degree to which the process meets DFE preferences for this matrix element.

- Is the process designed to minimize the use of materials whose extraction or purification involves the generation of large amounts of liquid residues? (This category includes paper and allied products, coal, and materials from biomass.)

- Is the process designed to minimize the use of materials whose extraction or purification involves the generation of toxic liquid residues? (These materials are defined as aluminum, chromium, copper, iron, lead, mercury, nickel, and zinc.)
- Are refillable/reusable containers used for incoming liquid materials where appropriate?
- Does the use of incoming components require cleaning that involves a large amount of water or that generates liquid residues needing special disposal methods?

PROCESS MATRIX ELEMENT: 1,5

Life Stage: Resource Provisioning
Environmental Stressor: Gaseous Residues

If the following condition applies, the matrix element rating is 0:

- The materials used cause substantial emissions of toxic, smog-producing, or greenhouse gases into the environment, and suitable alternatives that do not do so are available. (These materials are defined as aluminum, copper, iron, lead, nickel, zinc, paper and allied products, and concrete.)

If the following condition applies, the matrix element rating is 4:

- No gaseous residues are produced during resource extraction or production of materials by recycling.

If neither of the preceding ratings is assigned, complete the checklist below. Assign a rating of 1, 2, or 3 depending on the degree to which the process meets DFE preferences for this matrix element.

- Is the process designed to minimize the use of materials whose extraction or purification involves the generation of large amounts of gaseous (toxic or otherwise) residues? (Such materials are defined as aluminum, copper, iron, lead, nickel, and zinc.)
- Does the process design avoid using consumable materials whose transport to the facility will result in significant gaseous residues?

PROCESS MATRIX ELEMENT: 2,1

Life Stage: Process Implementation
Environmental Stressor: Materials Choice

If the following condition applies, the matrix element rating is 0:

- Process equipment manufacture requires relatively large amounts of materials that are restricted, toxic, and/or radioactive.

If the following condition applies, the matrix element rating is 4:

- Materials used in process equipment manufacture are completely closed loop (captured and reused/recycled) with minimum inputs required.

If neither of the preceding ratings is assigned, complete the checklist below. Assign a rating of 1, 2, or 3 depending on the degree to which the process equipment meets DFE preferences for this matrix element.

- Does the process equipment design avoid the use of materials that are in restricted supply?
- Is the use of toxic material avoided or minimized?
- Is the use of radioactive material avoided?
- Is the use of virgin material minimized?
- Has the chemical treatment of materials and components been minimized?

PROCESS MATRIX ELEMENT: 2,2

Life Stage: Process Implementation
Environmental Stressor: Energy Use

If the following condition applies, the matrix element rating is 0:

- Energy use for process equipment manufacture and installation is high and less energy-intensive alternatives are available.

If the following condition applies, the matrix element rating is 4:

- Process equipment manufacture and installation requires no or minimal energy use.

If neither of the preceding ratings is assigned, complete the checklist below. Assign a rating of 1, 2, or 3 depending on the degree to which the process equipment meets DFE preferences for this matrix element.

- Is the process equipment manufacture designed to minimize the use of energy-intensive processing steps?
- Is the process equipment manufacture designed to minimize energy-intensive evaluation/testing steps?
- Does the process equipment manufacture use co-generation, heat exchanges, and/or other techniques to utilize otherwise wasted energy?

PROCESS MATRIX ELEMENT: 2,3

Life Stage: Process Implementation
Environmental Stressor: Solid Residues

If the following condition applies, the matrix element rating is 0:

- Solid residues from shipping and installation are large and no reuse/recycling programs are in use.

If the following condition applies, the matrix element rating is 4:

- Solid residues from shipping and installation are minor and are >90 percent reused/recycled.

If neither of the preceding ratings is assigned, complete the checklist below. Assign a rating of 1, 2, or 3 depending on the degree to which the process equipment meets DFE preferences for this matrix element.

- Have solid shipping and installation residues been minimized to the greatest extent possible?
- Has the resale of all solid shipping and installation residues been implemented?

PROCESS MATRIX ELEMENT: 2,4

Life Stage: Process Implementation
Environmental Stressor: Liquid Residues

If the following condition applies, the matrix element rating is 0:

- Liquid shipping and installation residues are large and no reuse/recycling programs are in use.

If the following condition applies, the matrix element rating is 4:

- Liquid shipping and installation residues are minor and each constituent is >90 percent reused/recycled.

If neither of the preceding ratings is assigned, complete the checklist below. Assign a rating of 1, 2, or 3 depending on the degree to which the process meets DFE preferences for this matrix element.

- If solvents or oils are used in the shipping and installation of this process equipment, is their use minimized and have alternatives been investigated and implemented?
- If water is used in the shipping and installation of this process equipment, is its use minimized and have alternative approaches been investigated and implemented?

PROCESS MATRIX ELEMENT: 2,5

Life Stage: Process Implementation
Environmental Stressor: Gaseous Residues

If either of the following conditions apply, the matrix element rating is 0:

- Gaseous process equipment shipping and installation residues are large and uncontrolled.

- CFCs are used in process equipment shipping and installation.

  If the following condition applies, the matrix element rating is 4:

- Gaseous process equipment shipping and installation residues are relatively minor and reuse/recycling programs are in use.

  If neither of the preceding ratings is assigned, complete the checklist below. Assign a rating of 1, 2, or 3 depending on the degree to which the process meets DFE preferences for this matrix element.

  - Are greenhouse gases used or generated in any shipping or installation activity connected with the process equipment? Have the emitted amounts been minimized?

PROCESS MATRIX ELEMENT: 3,1

Life Stage: Primary Process Operation
Environmental Stressor: Materials Choice

If the following condition applies, the matrix element rating is 0:

- Large quantities of toxic and/or scarce consumables are used in the primary process.

If the following condition applies, the matrix element rating is 4:

- No toxic and/or scarce consumables are used in the primary process.

If neither of the preceding ratings is assigned, complete the checklist below. Assign a rating of 1, 2, or 3 depending on the degree to which the primary process meets DFE preferences for this matrix element.

- Is the use of toxic consumable materials in primary processes avoided or minimized?
- Is the use of radioactive consumable materials in primary processes avoided or minimized?
- Is the primary process designed to avoid the use of large amounts of water?

PROCESS MATRIX ELEMENT: 3,2

Life Stage: Primary Process Operation
Environmental Stressor: Energy Use

If the following condition applies, the matrix element rating is 0:

- Energy use in the primary process is very high and no conservation is practiced.

If the following condition applies, the matrix element rating is 4:

- Negligible energy is used in the primary manufacturing process.

If neither of the preceding ratings is assigned, complete the checklist below. Assign a rating of 1, 2, or 3 depending on the degree to which the primary process meets DFE preferences for this matrix element.

- Does the primary process use energy-efficient equipment such as variable-speed motors?
- Is the primary process designed to minimize the use of energy-intensive process steps such as high heating differentials, heavy motors, extensive cooling, and so on?
- Is the primary process designed to minimize the use of energy-intensive evaluation steps such as testing in a heated chamber?
- Does the primary process use co-generation, heat exchange, and other techniques for utilizing otherwise wasted energy?

PROCESS MATRIX ELEMENT: 3,3

Life Stage: Primary Process Operation
Environmental Stressor: Solid Residues

If the following condition applies, the matrix element rating is 0:

- Large amounts of solid residues result from the primary process and no control is practiced.

If the following condition applies, the matrix element rating is 4:

- No solid residues result from the primary process.

If neither of the preceding ratings is assigned, complete the checklist below. Assign a rating of 1, 2, or 3 depending on the degree to which the primary process meets DFE preferences for this matrix element.

- Have solid primary process residues (mold scrap, cutting scrap, and so on) been minimized and reused to the greatest extent possible?
- Have opportunities for sale of all primary process solid residues as inputs into the products and processes of others been investigated, and modifications made to residues (if possible and necessary) to facilitate such transactions?
- Has packaging material entering the facility in connection with primary processes been minimized, and does it use the fewest possible different materials?
- Do suppliers take back the packaging material in which consumables for primary processes enter the facility?

PROCESS MATRIX ELEMENT: 3,4

Life Stage: Primary Process Operation
Environmental Stressor: Liquid Residues

If the following condition applies, the matrix element rating is 0:

- Large amounts of liquid residues result from the primary process and no control is practiced.

If the following condition applies, the matrix element rating is 4:

- No liquid residues result from the primary process.

If neither of the preceding ratings is assigned, complete the checklist below. Assign a rating of 1, 2, or 3 depending on the degree to which the primary process meets DFE preferences for this matrix element.

- If solvents or oils are used in the primary process, is their use minimized and have substitutes been investigated?
- Have opportunities for sale of all primary process liquid residues as inputs into the products and processes of others been investigated, and modifications made to residues (if possible and necessary) to facilitate such transactions?
- Has the primary process been designed to utilize the maximum amount of recycled liquid species rather than virgin materials?

PROCESS MATRIX ELEMENT: 3,5

Life Stage: Primary Process Operation
Environmental Stressor: Gaseous Residues

If the following condition applies, the matrix element rating is 0:

- Large amounts of gaseous residues result from the primary process and no control is practiced.

If the following condition applies, the matrix element rating is 4:

- No gaseous residues result from the primary process.

If neither of the preceding ratings is assigned, complete the checklist below. Assign a rating of 1, 2, or 3 depending on the degree to which the primary process meets DFE preferences for this matrix element.

- Have opportunities for sale of all primary process gaseous residues as inputs into the products and processes of others been investigated, and modifications made to residues (if possible and necessary) to facilitate such transactions?
- Are greenhouse gases used or generated in the primary process?
- If CFCs or HCFCs are used in the primary process, have alternatives been thoroughly investigated?

PROCESS MATRIX ELEMENT: 4,1

Life Stage: Complementary Process Operation
Environmental Stressor: Materials Choice

If the following condition applies, the matrix element rating is 0:

- Large quantities of toxic and/or scarce consumables are used in complementary processes.

Appendix B   Environmentally Responsible Process Matrix   259

If the following condition applies, the matrix element rating is 4:

- No toxic and/or scarce consumables are used in complementary processes.

If neither of the preceding ratings is assigned, complete the checklist below. Assign a rating of 1, 2, or 3 depending on the degree to which complementary processes meet DFE preferences for this matrix element.

- Is the use of toxic consumable materials in complementary processes avoided or minimized?
- Is the use of radioactive consumable materials in complementary processes avoided or minimized?
- Are complementary processes designed to avoid the use of large amounts of water?

PROCESS MATRIX ELEMENT: 4,2

Life Stage: Complementary Process Operation
Environmental Stressor: Energy Use

If the following condition applies, the matrix element rating is 0:

- Energy use in complementary processes is very high and no conservation is practiced.

If the following condition applies, the matrix element rating is 4:

- Negligible energy is used in the complementary processes.

If neither of the preceding ratings is assigned, complete the checklist below. Assign a rating of 1, 2, or 3 depending on the degree to which the complementary processes meet DFE preferences for this matrix element.

- Do the complementary processes use energy-efficient equipment such as variable-speed motors?
- Are complementary processes designed to minimize the use of energy-intensive process steps such as high heating differentials, heavy motors, extensive cooling, and so on?
- Are complementary processes designed to minimize the use of energy-intensive evaluation steps such as testing in a heated chamber?
- Do complementary processes use co-generation, heat exchange, and other techniques for utilizing otherwise wasted energy?

PROCESS MATRIX ELEMENT: 4,3

Life Stage: Complementary Process Operation
Environmental Stressor: Solid Residues

If the following condition applies, the matrix element rating is 0:

- Large amounts of solid residues result from the complementary processes and no control is practiced.

If the following condition applies, the matrix element rating is 4:

- No solid residues result from the complementary processes.

If neither of the preceding ratings is assigned, complete the checklist below. Assign a rating of 1, 2, or 3 depending on the degree to which the complementary processes meet DFE preferences for this matrix element.

- Have solid complementary process residues (mold scrap, cutting scrap, and so on) been minimized and reused to the greatest extent possible?
- Have opportunities for sale of all complementary process solid residues as inputs into the products and processes of others been investigated, and modifications made to residues (if possible and necessary) to facilitate such transactions?
- Has packaging material entering the facility in connection with complementary processes been minimized, and does it use the fewest possible different materials?
- Do suppliers take back the packaging material in which consumables for complementary processes enter the facility?

PROCESS MATRIX ELEMENT: 4,4

Life Stage: Complementary Process Operation
Environmental Stressor: Liquid Residues

If the following condition applies, the matrix element rating is 0:

- Large amounts of liquid residues result from the complementary processes and no control is practiced.

If the following condition applies, the matrix element rating is 4:

- No liquid residues result from the complementary processes.

If neither of the preceding ratings is assigned, complete the checklist below. Assign a rating of 1, 2, or 3 depending on the degree to which the complementary processes meet DFE preferences for this matrix element.

- If solvents or oils are used in complementary processes, is their use minimized and have substitutes been investigated?
- Have opportunities for sale of all complementary process liquid residues as inputs into the products and processes of others been investigated, and modifications made to residues (if possible and necessary) to facilitate such transactions?
- Have complementary processes been designed to utilize the maximum amount of recycled liquid species rather than virgin materials?

PROCESS MATRIX ELEMENT: 4,5

Life Stage: Complementary Process Operation
Environmental Stressor: Gaseous Residues

If the following condition applies, the matrix element rating is 0:

- Large amounts of gaseous residues result from the complementary processes and no control is practiced.

If the following condition applies, the matrix element rating is 4:

- No gaseous residues result from the complementary processes.

If neither of the preceding ratings is assigned, complete the checklist below. Assign a rating of 1, 2, or 3 depending on the degree to which the complementary processes meet DFE preferences for this matrix element.

- Have opportunities for sale of all complementary process gaseous residues as inputs into the products and processes of others been investigated, and modifications made to residues (if possible and necessary) to facilitate such transactions?
- Are greenhouse gases used or generated in the complementary processes?
- If CFCs or HCFCs are used in complementary processes, have alternatives been thoroughly investigated?

PROCESS MATRIX ELEMENT: 5,1

Life Stage: Refurbishment, Recycling, Disposal
Environmental Stressor: Material Choice

If the following condition applies, the matrix element rating is 0:

- Process equipment contains significant quantities of mercury (i.e., mercury relays), asbestos (i.e., asbestos based insulations), or cadmium (i.e., cadmium or zinc plated parts) that are not clearly identified and easily removable.

If the following condition applies, the matrix element rating is 4:

- Material diversity is minimized, the product is easy to disassemble, and all parts are recyclable.

If neither of the preceding ratings is assigned, complete the checklist below. Assign a rating of 1, 2, or 3 depending on the degree to which the process equipment meets DFE preferences for this matrix element.

- Have materials been chosen and used in light of the desired recycling/disposal option for the process equipment (e.g., for incineration, for recycling, for refurbishment)?
- Does the process equipment minimize the number of different materials that are used in its manufacture?
- Are the different materials easy to identify and separate?

- Is this battery-free process equipment?
- Is this process equipment free of components containing PCBs or PCTs (e.g., in capacitors and transformers)?
- Are major plastics parts free of polybrominated flame retardants or heavy metal-based additives (colorants, conductors, stabilizers, etc.)?

PROCESS MATRIX ELEMENT: 5,2

Life Stage: Refurbishment, Recycling, Disposal
Environmental Stressor: Energy Use

If the following condition applies, the matrix element rating is 0:

- Recycling/disposal of this process equipment is relatively energy intensive (compared to other products that perform the same function) due to its weight, construction, and/or complexity.

If the following condition applies, the matrix element rating is 4:

- Energy use for recycling or disposal of this process equipment is minimal.

If neither of the preceding ratings is assigned, complete the checklist below. Assign a rating of 1, 2, or 3 depending on the degree to which the process equipment meets DFE preferences for this matrix element.

- Is the process equipment designed with the aim of minimizing the use of energy-intensive steps in disassembly?
- Is the process equipment designed for high-level reuse of materials? (Direct reuse in similar process equipment is preferable to a degraded reuse.)
- Will transport of process equipment for recycling be energy intensive because of process equipment weight or volume or the location of recycling facilities?

PROCESS MATRIX ELEMENT: 5,3

Life Stage: Refurbishment, Recycling, Disposal
Environmental Stressor: Solid Residues

If the following condition applies, the matrix element rating is 0:

- Process equipment consists primarily of unrecyclable solid materials (such as rubber, fiberglass, and compound polymers).

If the following condition applies, the matrix element rating is 4:

- Process equipment can be easily refurbished and reused and is easily dismantled and 100 percent reused/recycled at the end of its life. For example, no part of this process equipment will end up in a landfill.

If neither of the preceding ratings is assigned, complete the checklist below. Assign a rating of 1, 2, or 3 depending on the degree to which the process equipment meets DFE preferences for this matrix element.

- Has the process equipment been assembled with fasteners such as clips or hook-and-loop attachments rather than chemical bonds (gels, potting compounds) or welds?
- Have efforts been made to avoid joining dissimilar materials together in ways difficult to reverse?
- Are all plastic components identified by ISO markings as to their content?
- If the process equipment consists of plastic parts, is there one dominant (>80 percent by weight) species?
- Is this process equipment to be leased rather than sold?

PROCESS MATRIX ELEMENT: 5,4

Life Stage: Refurbishment, Recycling, Disposal
Environmental Stressor: Liquid Residues

If the following condition applies, the matrix element rating is 0:

- The process equipment contains primarily unrecyclable liquid materials.

If the following condition applies, the matrix element rating is 4:

- The process equipment uses no operating liquids (such as oils, coolants, or hydraulic fluids) and no cleaning agents or solvents are necessary for its reconditioning.

If neither of the preceding ratings is assigned, complete the checklist below. Assign a rating of 1, 2, or 3 depending on the degree to which the process equipment meets DFE preferences for this matrix element.

- Can liquids contained in the process equipment be recovered at disassembly rather than lost?
- Does disassembly, recovery, and reuse generate liquid residues?
- Does materials recovery and reuse generate liquid residues?

PROCESS MATRIX ELEMENT: 5,5

Life Stage: Refurbishment, Recycling, Disposal
Environmental Stressor: Gaseous Residues

If the following condition applies, the matrix element rating is 0:

- Process equipment contains or produces primarily unrecyclable gaseous materials that are dissipated to the atmosphere at the end of its life.

If the following condition applies, the matrix element rating is 4:

- Process equipment contains no substances lost to evaporation/ sublimation (other than water) and no volatile substances are used for refurbishment.

If neither of the preceding ratings is assigned, complete the checklist below. Assign a rating of 1, 2, or 3 depending on the degree to which the process equipment meets DFE preferences for this matrix element.

- Can gases contained in the process equipment be recovered at disassembly rather than lost?
- Does materials recovery and reuse generate gaseous residues?
- Can plastic parts be incinerated without requiring sophisticated air pollution control devices? Plastic parts that can cause difficulty in this regard are those that contain polybrominated flame retardants or metal-based additives, are finished with polyurethane-based paints, or are plated or painted with metals.

APPENDIX C

Environmentally Responsible Facilities Matrix: Scoring Guidelines and Protocols

The Facilities Improvement Matrix is described in Chapter 9. In this appendix, a sample of possible scoring considerations appropriate to each of the matrix elements is presented. It is anticipated that different types of facilities may require different checklists and evaluations, so this appendix is presented as an example rather than as a universal formula.

FACILITY MATRIX ELEMENT: 1,1

Life Stage: Site Selection, Development, and Infrastructure
Environmental Stressor: Ecological Impacts

If the following condition applies, the matrix element rating is 0:

- The site is developed with massive disturbance or destruction of natural areas.

If the following condition applies, the matrix element rating is 4:

- The site is developed with negligible destruction of natural areas.

If neither of the preceding ratings is assigned, complete the checklist below. Assign a rating of 1, 2, or 3 depending on the degree to which the facility meets DFE preferences for this matrix element.

- Has the proposed site previously been used for similar activities?
- If not, have any such sites been surveyed for availability?
- Is necessary development activity, if any, planned to avoid disruption of existing biological communities?
- Is the use of heavy equipment for site development minimized?
- Are areas that are disturbed during site development carefully restored?
- Is the biota of the site compatible with all planned facility emissions, including possible exceedances?

FACILITY MATRIX ELEMENT: 1,2

Life Stage: Site Selection, Development, and Infrastructure
Environmental Stressor: Energy Use

If the following condition applies, the matrix element rating is 0:

- A complete new energy infrastructure is installed during site development.

If the following condition applies, the matrix element rating is 4:

- The existing energy infrastructure is used without modification.

If neither of the preceding ratings is assigned, complete the checklist below. Assign a rating of 1, 2, or 3 depending on the degree to which the facility meets DFE preferences for this matrix element.

- Is the site such that it can be made operational with only minimal energy expenditures?
- Has the site been selected so as to avoid any energy emission impacts on existing biota?
- Does the site enable delivery and installation of construction or renovation materials with minimal use of energy?

FACILITY MATRIX ELEMENT: 1,3

Life Stage: Site Selection, Development, and Infrastructure
Environmental Stressor: Solid Residues

If the following condition applies, the matrix element rating is 0:

- A large quantity of solid residues is generated during site development and no recycling is practiced.

If the following condition applies, the matrix element rating is 4:

- Insignificant solid residues are generated during site development.

If neither of the preceding ratings is assigned, complete the checklist below. Assign a rating of 1, 2, or 3 depending on the degree to which the facility meets DFE preferences for this matrix element.

- Is the site such that it can be made operational with only minimal production of solid residues?
- Have plans been made to ensure that any solid residues generated in the process of developing the site are managed so as to minimize their impacts on biota and human health?
- If any solid residues generated in the process of developing the site may be hazardous or toxic to biota or humans, have plans been made to minimize releases and exposures?

FACILITY MATRIX ELEMENT: 1,4

Life Stage: Site Selection, Development, and Infrastructure
Environmental Stressor: Liquid Residues

If the following condition applies, the matrix element rating is 0:

- A large quantity of liquid residues is generated during site development and no recycling is practiced.

If the following condition applies, the matrix element rating is 4:

- Insignificant liquid residues are generated during site development.

If neither of the preceding ratings is assigned, complete the checklist below. Assign a rating of 1, 2, or 3 depending on the degree to which the facility meets DFE preferences for this matrix element.

- Is the site such that it can be made operational with only minimal production of liquid residues?
- Have plans been made to ensure that any liquid residues generated in the process of developing the site are managed so as to minimize their impacts on biota and human health?
- If any liquid residues generated in the process of developing the site may be hazardous or toxic to biota or humans, have plans been made to minimize releases and exposures?

FACILITY MATRIX ELEMENT: 1,5

Life Stage: Site Selection, Development, and Infrastructure
Environmental Stressor: Gaseous Residues

If the following condition applies, the matrix element rating is 0:

- A large quantity of gaseous residues is generated during site development and no recycling is practiced.

If the following condition applies, the matrix element rating is 4:

- Insignificant gaseous residues are generated during site development.

If neither of the preceding ratings is assigned, complete the checklist below. Assign a rating of 1, 2, or 3 depending on the degree to which the facility meets DFE preferences for this matrix element.

- Is the site such that it can be made operational with only minimal production of gaseous residues?
- Have plans been made to ensure that any gaseous residues generated in the process of developing the site are managed so as to minimize their impacts on biota and human health?

- If any gaseous residues generated in the process of developing the site may be hazardous or toxic to biota or humans, have plans been made to minimize releases and exposures?

FACILITY MATRIX ELEMENT: 2,1

Life Stage: Principal Business Activity—Products
Environmental Stressor: Ecological Impacts

If the following condition applies, the matrix element rating is 0:

- Hazardous and/or virgin materials are used where suitable nonhazardous and/or recycled materials exist.

If the following condition applies, the matrix element rating is 4:

- Negligible amounts of hazardous and/or virgin materials are used in the facilities equipment.

If neither of the preceding ratings is assigned, complete the checklist below. Assign a rating of 1, 2, or 3 depending on the degree to which the facility meets DFE preferences for this matrix element.

- Does the process development avoid the use of materials that are in restricted supply?
- Is the use of toxic material avoided or minimized?
- Is the use of radioactive material avoided?
- Is the use of virgin material minimized?
- Has the chemical treatment of materials and components been minimized?

FACILITY MATRIX ELEMENT: 2,2

Life Stage: Principal Business Activity—Products
Environmental Stressor: Energy Use

If the following condition applies, the matrix element rating is 0:

- Energy consumption for facility operations is very high and no energy management is practiced.

If the following condition applies, the matrix element rating is 4:

- Negligible energy consumption is involved in facility operations.

If neither of the preceding ratings is assigned, complete the checklist below. Assign a rating of 1, 2, or 3 depending on the degree to which the facility meets DFE preferences for this matrix element.

- Is the product and process designed to minimize the use of energy- intensive processing steps?

- Is the product and process designed to minimize energy-intensive evaluation/testing steps?
- Does the process use co-generation, heat exchanges, and/or other techniques to utilize otherwise wasted energy?
- Is the manufacturing facility powered down when not in use?

FACILITY MATRIX ELEMENT: 2,3

Life Stage: Principal Business Activity—Products
Environmental Stressor: Solid Residues

If the following condition applies, the matrix element rating is 0:

- Large amounts of solid residues are produced by facility operations equipment manufacture and installation.

If the following condition applies, the matrix element rating is 4:

- Negligible amounts of solid residues are produced by product manufacture.

If neither of the preceding ratings is assigned, complete the checklist below. Assign a rating of 1, 2, or 3 depending on the degree to which the facility meets DFE preferences for this matrix element.

- Have solid manufacturing residues been minimized and reused to the greatest extent possible?
- Has the resale of all solid residues as inputs to other products/processes been investigated and implemented?
- Are solid manufacturing residues that do not have resale value minimized and recycled?

FACILITY MATRIX ELEMENT: 2,4

Life Stage: Principal Business Activity—Products
Environmental Stressor: Liquid Residues

If the following condition applies, the matrix element rating is 0:

- Large amounts of liquid residues are produced by product manufacture and installation.

If the following condition applies, the matrix element rating is 4:

- Negligible amounts of liquid residues are produced by product manufacture and installation.

If neither of the preceding ratings is assigned, complete the checklist below. Assign a rating of 1, 2, or 3 depending on the degree to which the facility meets DFE preferences for this matrix element.

- If solvents or oils are used in the manufacture of this product, is their use minimized and have alternatives been investigated and implemented?
- Have opportunities for sale of all liquid residues as input to other processes/products been investigated and implemented?
- Have the processes been designed to require the maximum recycled liquid process chemicals rather than virgin materials?

FACILITY MATRIX ELEMENT: 2,5

Life Stage: Principal Business Activity—Products
Environmental Stressor: Gaseous Residues

If the following condition applies, the matrix element rating is 0:

- Large amounts of gaseous residues are produced by product manufacture and installation.

If the following condition applies, the matrix element rating is 4:

- Negligible amounts of gaseous residues are produced by product manufacture and installation.

If neither of the preceding ratings is assigned, complete the checklist below. Assign a rating of 1, 2, or 3 depending on the degree to which the facility meets DFE preferences for this matrix element.

- If HCFCs are used in the manufacture of this product, have alternatives been thoroughly investigated and implemented?
- Are greenhouse gases used or generated in any manufacturing process connected with this product?
- Have the resale of all gaseous residues as inputs to other processes/products been investigated and implemented?

FACILITY MATRIX ELEMENT: 3,1

Life Stage: Principal Business Activity—Processes
Environmental Stressor: Ecological Impacts

If the following condition applies, the matrix element rating is 0:

- Large quantities of toxic and/or scarce consumables are used in the process.

If the following condition applies, the matrix element rating is 4:

- No toxic and/or scarce consumables are used in facility processes.

If neither of the preceding ratings is assigned, complete the checklist below. Assign a rating of 1, 2, or 3 depending on the degree to which the facility meets DFE preferences for this matrix element.

Appendix C   Environmentally Responsible Facilities Matrix   271

- Is the use of toxic consumables materials in manufacturing processes avoided or minimized?
- Is the use of radioactive materials in manufacturing processes avoided or minimized?
- Is the use of water in manufacturing processes avoided or minimized?

FACILITY MATRIX ELEMENT: 3,2

Life Stage: Principal Business Activity—Processes
Environmental Stressor: Energy Use

If the following condition applies, the matrix element rating is 0:

- Energy used in manufacturing processes is very high and no conservation is practiced.

If the following condition applies, the matrix element rating is 4:

- Negligible energy is used in manufacturing operations.

If neither of the preceding ratings is assigned, complete the checklist below. Assign a rating of 1, 2, or 3 depending on the degree to which the facility meets DFE preferences for this matrix element.

- Is vehicular activity in connection with manufacturing activities minimized?
- Does the manufacturing involve the use of energy-intensive activities such as high heating differentials, heavy motors, extensive cooling, and so on, and have these activities been minimized as much as possible?

FACILITY MATRIX ELEMENT: 3,3

Life Stage: Principal Business Activity—Processes
Environmental Stressor: Solid Residues

If the following condition applies, the matrix element rating is 0:

- Large amounts of solid residues result from the manufacturing activities and no control is practiced.

If the following condition applies, the matrix element rating is 4:

- No solid residues result from the manufacturing activities.

If neither of the preceding ratings is assigned, complete the checklist below. Assign a rating of 1, 2, or 3 depending on the degree to which the facility meets DFE preferences for this matrix element.

- Has packaging material from suppliers of replacement parts or consumables been minimized, and does it use the fewest possible different materials? Is it made from recycled materials?

- Are manufacturing activities designed so as to generate minimal and nontoxic solid residues?
- Are solid residues from manufacturing activities recycled?

FACILITY MATRIX ELEMENT: 3,4

Life Stage: Principal Business Activity—Processes
Environmental Stressor: Liquid Residues

If the following condition applies, the matrix element rating is 0:

- Large amounts of liquid residues result from the manufacturing activities and no control is practiced.

If the following condition applies, the matrix element rating is 4:

- No liquid residues result from the manufacturing activities.

If neither of the preceding ratings is assigned, complete the checklist that follows. Assign a rating of 1, 2, or 3 depending on the degree to which the facility meets DFE preferences for this matrix element.

- Have manufacturing activities been designed to utilize the maximum amount of recycled liquid consumables rather than virgin materials?
- Are manufacturing activities designed to generate minimal and nontoxic liquid residues in use?
- Are liquid residues from manufacturing activities recycled?

FACILITY MATRIX ELEMENT: 3,5

Life Stage: Principal Business Activity—Processes
Environmental Stressor: Gaseous Residues

If the following condition applies, the matrix element rating is 0:

- Large amounts of gaseous residues result from the manufacturing activities and no control is practiced.

If the following condition applies, the matrix element rating is 4:

- No gaseous residues result from the manufacturing activities.

If neither of the preceding ratings is assigned, complete the checklist below. Assign a rating of 1, 2, or 3 depending on the degree to which the facility meets DFE preferences for this matrix element.

- Are manufacturing activities designed to generate minimal and nontoxic gaseous residues in use?
- If CFCs or HCFCs are used in the manufacturing activities, have alternatives been thoroughly investigated?

- Are greenhouse gases used or generated in processes within the facility?
- Are products designed to generate minimal and nontoxic gaseous residues in recycling or disposal?
- Are gaseous residues from manufacturing activities recycled.

FACILITY MATRIX ELEMENT: 4,1

Life Stage: Facility Operations
Environmental Stressor: Ecological Impacts

If the following condition applies, the matrix element rating is 0:

- None of the site is left in a natural state and fertilizers and pesticides are used freely.

If the following condition applies, the matrix element rating is 4:

- Essentially all the site has been left in its natural state except for immediate building areas, and no pesticides or fertilizer are used.

If neither of the preceding ratings is assigned, complete the checklist below. Assign a rating of 1, 2, or 3 depending on the degree to which the facility meets DFE preferences for this matrix element.

- Is the maximum possible portion of the facility allowed to remain in its natural state?
- Is the use of pesticides and herbicides on the property minimized?
- Is noise pollution from the site minimized?

FACILITY MATRIX ELEMENT: 4,2

Life Stage: Facility Operations
Environmental Stressor: Energy Use

If the following condition applies, the matrix element rating is 0:

- No site energy management of any kind is practiced.

If the following condition applies, the matrix element rating is 4:

- Energy use for facility operations is minimal and is provided by renewable resources (solar, wind power, and so on).

If neither of the preceding ratings is assigned, complete the checklist below. Assign a rating of 1, 2, or 3 depending on the degree to which the facility meets DFE preferences for this matrix element.

- Is the energy needed for heating and cooling the buildings minimized?
- Is the energy needed for lighting the buildings minimized?

- Is energy efficiency a consideration when buying or leasing facility equipment: copiers, computers, fan motors, and the like?

FACILITY MATRIX ELEMENT: 4,3

Life Stage: Facility Operations
Environmental Stressor: Solid Residues

If the following condition applies, the matrix element rating is 0:

- Solid residues from facility operations are large and are given the minimum treatment necessary to comply with legal requirements.

If the following condition applies, the matrix element rating is 4:

- Solid residues from facility operations are negligible or are completely recycled on site.

If neither of the preceding ratings is assigned, complete the checklist below. Assign a rating of 1, 2, or 3 depending on the degree to which the facility meets DFE preferences for this matrix element.

- Is the facility designed to minimize the co-mingling of solid waste streams?
- Are solid residues from facility operations reused or recycled to the extent possible?
- Are unusable solid residues from facility operations (including food service) disposed of in an environmentally responsible manner?

FACILITY MATRIX ELEMENT: 4,4

Life Stage: Facility Operations
Environmental Stressor: Liquid Residues

If the following condition applies, the matrix element rating is 0:

- Liquid residues from facility operations are large and are given the minimum treatment necessary to comply with legal requirements.

If the following condition applies, the matrix element rating is 4:

- Liquid residues (including stormwater) from facility operations are negligible or are completely recycled on site.

If neither of the preceding ratings is assigned, complete the checklist below. Assign a rating of 1, 2, or 3 depending on the degree to which the facility meets DFE preferences for this matrix element.

- Is the facility designed to minimize the co-mingling of liquid waste streams?
- Are liquid treatment plants monitored to ensure that they operate at peak efficiency?

- Are unusable liquid residues from facility operations disposed of in an environmentally responsible manner?

FACILITY MATRIX ELEMENT: 4,5

Life Stage: Facility Operations
Environmental Stressor: Gaseous Residues

If the following condition applies, the matrix element rating is 0:

- Gaseous residues from facility operations are large and are given the minimum treatment necessary to comply with legal requirements.

If the following condition applies, the matrix element rating is 4:

- Gaseous residues from facility operations are negligible or are completely recycled on site.

If neither of the preceding ratings is assigned, complete the checklist below. Assign a rating of 1, 2, or 3 depending on the degree to which the facility meets DFE preferences for this matrix element.

- Is facility operations-related transportation to and from the facility minimized?
- Are furnaces, incinerators, and other combustion processes and their related air pollution control devices monitored to ensure operation at peak efficiency?
- Is employee commuting minimized by job sharing, telecommuting, and similar programs?

FACILITY MATRIX ELEMENT: 5,1

Life Stage: Facility Refurbishment, Transfer, and Closure
Environmental Stressor: Ecological Impacts

If the following condition applies, the matrix element rating is 0:

- Structures must be completely demolished, with major impacts on natural areas or existing ecosystems.

If the following condition applies, the matrix element rating is 4:

- Site and structures can be reused with negligible impact on natural areas or existing ecosystems.

If neither of the preceding ratings is assigned, complete the checklist below. Assign a rating of 1, 2, or 3 depending on the degree to which the facility meets DFE preferences for this matrix element.

- Will site and system closure have substantial impacts on natural areas?
- Will site and system closure have substantial impacts on existing ecosystems?

FACILITY MATRIX ELEMENT: 5,2

Life Stage: Facility Refurbishment, Transfer, and Closure
Environmental Stressor: Energy Use

If the following condition applies, the matrix element rating is 0:

- Very large energy consumption will be required for facility closure.

If the following condition applies, the matrix element rating is 4:

- Negligible energy consumption will be required for facility closure.

If neither of the preceding ratings is assigned, complete the checklist below. Assign a rating of 1, 2, or 3 depending on the degree to which the facility meets DFE preferences for this matrix element.

- Will site closure require the transport of large quantities of materials?
- Will building demolition or renovation require large energy expenditures?
- Will equipment demolition or renovation require large energy expenditures?

FACILITY MATRIX ELEMENT: 5,3

Life Stage: Facility Refurbishment, Transfer, and Closure
Environmental Stressor: Solid Residues

If the following condition applies, the matrix element rating is 0:

- Large amounts of solid residues will be produced by facility closure and avoidance is difficult or impossible.

If the following condition applies, the matrix element rating is 4:

- Negligible amounts of solid residues, and none with biotoxic characteristics, will be produced by facility closure.

If neither of the preceding ratings is assigned, complete the checklist below. Assign a rating of 1, 2, or 3 depending on the degree to which the facility meets DFE preferences for this matrix element.

- Does the facility contain asbestos or lead paint?
- Can facility equipment be reused or recycled?
- Can unwanted building fixtures and components be reused or recycled?
- Can building structural components be reused or recycled?
- Can spare parts, supplies, or other solid facility activity consumables or residues be used or recycled?

FACILITY MATRIX ELEMENT: 5,4

Life Stage: Facility Refurbishment, Transfer, and Closure
Environmental Stressor: Liquid Residues

If the following condition applies, the matrix element rating is 0:

- Large amounts of liquid residues will be produced by facility closure and avoidance is difficult or impossible.

If the following condition applies, the matrix element rating is 4:

- Negligible amounts of liquid residues will be produced by facility closure.

If neither of the preceding ratings is assigned, complete the checklist below. Assign a rating of 1, 2, or 3 depending on the degree to which the facility meets DFE preferences for this matrix element.

- Will liquid residues be produced during demolition?
- Will liquid residues be produced during facility termination?
- Can liquid facility activity consumables or residues be used or recycled?

FACILITY MATRIX ELEMENT: 5,5

Life Stage: Facility Refurbishment, Transfer, and Closure
Environmental Stressor: Gaseous Residues

If the following condition applies, the matrix element rating is 0:

- Large amounts of gaseous residues will be produced by facility closure and avoidance is difficult or impossible.

If the following condition applies, the matrix element rating is 4:

- Negligible amounts of gaseous residues will be produced by facility closure.

If neither of the preceding ratings is assigned, complete the checklist below. Assign a rating of 1, 2, or 3 depending on the degree to which the facility meets DFE preferences for this matrix element.

- Will gaseous residues be produced during demolition?
- Will gaseous residues be produced during termination of facilities operations?
- Can gaseous facility consumables or residues be used or recycled?

APPENDIX D: Environmentally Responsible Services Matrix: Scoring Guidelines and Protocols

The Service Improvement Matrix is described in Chapter 10. In this appendix, a sample of possible scoring considerations appropriate to each of the matrix elements is presented. It is anticipated that different types of services may require different check lists and evaluations, so this appendix is presented as an example rather than as a universal formula.

SERVICE MATRIX ELEMENT: 1,1

Life Stage: Site and Service Development
Environmental Stressor: Ecological Impacts

If the following condition applies, the matrix element rating is 0:

- The site is developed with massive disturbance or destruction of natural areas.

If the following condition applies, the matrix element rating is 4:

- The site is developed with negligible destruction of natural areas.

If neither of the preceding ratings is assigned, complete the checklist below. Assign a rating of 1, 2, or 3 depending on the degree to which the facility meets DFE preferences for this matrix element.

- Has the proposed site previously been used for similar activities? If not, have any such sites been surveyed for availability?
- Is necessary development activity, if any, planned to avoid disruption of existing biological communities?
- Is the use of heavy equipment for site development minimized?
- Are areas disturbed during site development carefully restored?
- Is the biota of the site compatible with all planned service activity emissions, including possible exceedances?

SERVICE MATRIX ELEMENT: 1,2

Life Stage: Site and Service Development
Environmental Stressor: Energy Use

If the following condition applies, the matrix element rating is 0:

- A complete new energy infrastructure is installed during site development.

If the following condition applies, the matrix element rating is 4:

- The existing energy infrastructure is used without modification.

If neither of the preceding ratings is assigned, complete the checklist below. Assign a rating of 1, 2, or 3 depending on the degree to which the facility meets DFE preferences for this matrix element.

- Is the site such that it can be made operational with only minimal energy expenditures?
- Has the site been selected so as to avoid any energy emission impacts on existing biota? Does the site enable delivery and installation of construction or renovation materials with minimal use of energy?

SERVICE MATRIX ELEMENT: 1,3

Life Stage: Site and Service Development
Environmental Stressor: Solid Residues

If the following condition applies, the matrix element rating is 0:

- A large quantity of solid residues is generated during site development and no recycling is practiced.

If the following condition applies, the matrix element rating is 4:

- Insignificant solid residues are generated during site development.

If neither of the preceding ratings is assigned, complete the checklist below. Assign a rating of 1, 2, or 3 depending on the degree to which the facility meets DFE preferences for this matrix element.

- Is the site such that it can be made operational with only minimal production of solid residues?
- Have plans been made to ensure that any solid residues generated in the process of developing the site are managed so as to minimize their impacts on biota and human health?
- If any solid residues generated in the process of developing the site may be hazardous or toxic to biota or humans, have plans been made to minimize releases and exposures?

SERVICE MATRIX ELEMENT: 1,4

Life Stage: Site and Service Development
Environmental Stressor: Liquid Residues

If the following condition applies, the matrix element rating is 0:

- A large quantity of liquid residues is generated during site development and no recycling is practiced.

If the following condition applies, the matrix element rating is 4:

- Insignificant liquid residues are generated during site development.

If neither of the preceding ratings is assigned, complete the checklist below. Assign a rating of 1, 2, or 3 depending on the degree to which the facility meets DFE preferences for this matrix element.

- Is the site such that it can be made operational with only minimal production of liquid residues?
- Have plans been made to ensure that any liquid residues generated in the process of developing the site are managed so as to minimize their impacts on biota and human health?
- If any liquid residues generated in the process of developing the site may be hazardous or toxic to biota or humans, have plans been made to minimize releases and exposures?

SERVICE MATRIX ELEMENT: 1,5

Life Stage: Site and Service Development
Environmental Stressor: Gaseous Residues

If the following condition applies, the matrix element rating is 0:

- A large quantity of gaseous residues is generated during site development and no recycling is practiced.

If the following condition applies, the matrix element rating is 4:

- Insignificant gaseous residues are generated during site development.

If neither of the preceding ratings is assigned, complete the checklist below. Assign a rating of 1, 2, or 3 depending on the degree to which the facility meets DFE preferences for this matrix element.

- Is the site such that it can be made operational with only minimal production of gaseous residues?
- Have plans been made to ensure that any gaseous residues generated in the process of developing the site are managed so as to minimize their impacts on biota and human health?

- If any gaseous residues generated in the process of developing the site may be hazardous or toxic to biota or humans, have plans been made to minimize releases and exposures?

SERVICE MATRIX ELEMENT: 2,1

Life Stage: Infrastructure Interactions
Environmental Stressor: Ecological Impacts

If the following condition applies, the matrix element rating is 0:

- Hazardous and/or virgin materials are used where suitable nonhazardous and/or recycled materials exist.

If the following condition applies, the matrix element rating is 4:

- Negligible amounts of hazardous and/or virgin materials are used in the service equipment.

If neither of the preceding ratings is assigned, complete the checklist below. Assign a rating of 1, 2, or 3 depending on the degree to which the facility meets DFE preferences for this matrix element.

- Does the process development avoid the use of materials that are in restricted supply?
- Is the use of toxic material avoided or minimized?
- Is the use of radioactive material avoided?
- Is the use of virgin material minimized?
- Has the chemical treatment of materials and components been minimized?

SERVICE MATRIX ELEMENT: 2,2

Life Stage: Infrastructure Interactions
Environmental Stressor: Energy Use

If the following condition applies, the matrix element rating is 0:

- Energy consumption for service provisioning is very high and no energy management is practiced.

If the following condition applies, the matrix element rating is 4:

- Negligible energy consumption is involved in service provisioning.

If neither of the preceding ratings is assigned, complete the checklist below. Assign a rating of 1, 2, or 3 depending on the degree to which the facility meets DFE preferences for this matrix element.

- Is the product and process designed to minimize the use of energy-intensive processing steps?

- Is the product and process designed to minimize energy-intensive evaluation/testing steps?
- Does the process use co-generation, heat exchanges, and/or other techniques to utilize otherwise wasted energy?
- Is the manufacturing facility powered down when not in use?

SERVICE MATRIX ELEMENT: 2,3

Life Stage: Infrastructure Interactions
Environmental Stressor: Solid Residues

If the following condition applies, the matrix element rating is 0:

- Large amounts of solid residues are produced by service equipment manufacture and installation.

If the following condition applies, the matrix element rating is 4:

- Negligible amounts of solid residues are produced by service equipment manufacture and installation.

If neither of the preceding ratings is assigned, complete the checklist below. Assign a rating of 1, 2, or 3 depending on the degree to which the facility meets DFE preferences for this matrix element.

- Have solid manufacturing residues been minimized and reused to the greatest extent possible?
- Has the resale of all solid residues as inputs to other products/processes been investigated and implemented?
- Are solid manufacturing residues that do not have resale value minimized and recycled?

SERVICE MATRIX ELEMENT: 2,4

Life Stage: Infrastructure Interactions
Environmental Stressor: Liquid Residues

If the following condition applies, the matrix element rating is 0:

- Large amounts of liquid residues are produced by service equipment manufacture and installation.

If the following condition applies, the matrix element rating is 4:

- Negligible amounts of liquid residues are produced by service equipment manufacture and installation.

Appendix D Environmentally Responsible Services Matrix   283

If neither of the preceding ratings is assigned, complete the checklist below. Assign a rating of 1, 2, or 3 depending on the degree to which the facility meets DFE preferences for this matrix element.

- If solvents or oils are used in the manufacture of this product, is their use minimized and have alternatives been investigated and implemented?
- Have opportunities for sale of all liquid residues as input to other processes/products been investigated and implemented?
- Have the processes been designed to require the maximum recycled liquid process chemicals rather than virgin materials?

SERVICE MATRIX ELEMENT: 2,5

Life Stage: Infrastructure Interactions
Environmental Stressor: Gaseous Residues

If the following condition applies, the matrix element rating is 0:

- Large amounts of gaseous residues are produced by service equipment manufacture and installation.

If the following condition applies, the matrix element rating is 4:

- Negligible amounts of gaseous residues are produced by service equipment manufacture and installation.

If neither of the preceding ratings is assigned, complete the checklist below. Assign a rating of 1, 2, or 3 depending on the degree to which the facility meets DFE preferences for this matrix element.

- If HCFCs are used in the manufacture of this product, have alternatives been thoroughly investigated and implemented?
- Are greenhouse gases used or generated in any manufacturing process connected with this product?
- Have the resale of all gaseous residues as inputs to other processes/products been investigated and implemented?

SERVICE MATRIX ELEMENT: 3,1

Life Stage: Providing the Service
Environmental Stressor: Ecological Impacts

If the following condition applies, the matrix element rating is 0:

- Large quantities of toxic and/or scarce consumables are used in service activities.

If the following condition applies, the matrix element rating is 4:

- No toxic and/or scarce consumables are used in service activities.

If neither of the preceding ratings is assigned, complete the checklist below. Assign a rating of 1, 2, or 3 depending on the degree to which the facility meets DFE preferences for this matrix element.

- Is the use of toxic consumables materials in providing the service avoided or minimized?
- Is the use of radioactive materials in providing the service avoided or minimized?
- Is the use of water in providing the service avoided or minimized?

SERVICE MATRIX ELEMENT: 3,2

Life Stage: Providing the Service
Environmental Stressor: Energy Use

If the following condition applies, the matrix element rating is 0:

- Energy used in service activities is very high and no conservation is practiced.

If the following condition applies, the matrix element rating is 4:

- Negligible energy is used in service activities.

If neither of the preceding ratings is assigned, complete the checklist below. Assign a rating of 1, 2, or 3 depending on the degree to which the facility meets DFE preferences for this matrix element.

- Is vehicular activity in connection with service activities minimized?
- Does the service involve the use of energy-intensive activities such as high heating differentials, heavy motors, extensive cooling, and so on, and have these activities been minimized as much as possible?

SERVICE MATRIX ELEMENT: 3,3

Life Stage: Providing the Service
Environmental Stressor: Solid Residues

If the following condition applies, the matrix element rating is 0:

- Large amounts of solid residues result from the service activities and no control is practiced.

If the following condition applies, the matrix element rating is 4:

- No solid residues result from the service activities.

If neither of the preceding ratings is assigned, complete the checklist below. Assign a rating of 1, 2, or 3 depending on the degree to which the facility meets DFE preferences for this matrix element.

- Has packaging material from suppliers of replacement parts or consumables been minimized, and does it use the fewest possible different materials?
- Is it made from recycled materials?
- Are service activities designed so as to generate minimal and nontoxic solid residues?
- Are solid residues from service activities recycled?

SERVICE MATRIX ELEMENT: 3,4

Life Stage: Providing the Service
Environmental Stressor: Liquid Residues

If the following condition applies, the matrix element rating is 0:

- Large amounts of liquid residues result from the service activities and no control is practiced.

If the following condition applies, the matrix element rating is 4:

- No liquid residues result from the service activities.

If neither of the preceding ratings is assigned, complete the checklist below. Assign a rating of 1, 2, or 3 depending on the degree to which the facility meets DFE preferences for this matrix element.

- Have service activities been designed to utilize the maximum amount of recycled liquid consumables rather than virgin materials?
- Are service activities designed to generate minimal and nontoxic liquid residues in use?
- Are liquid residues from service activities recycled?

SERVICE MATRIX ELEMENT: 3,5

Life Stage: Providing the Service
Environmental Stressor: Gaseous Residues

If the following condition applies, the matrix element rating is 0:

- Large amounts of gaseous residues result from the service activities and no control is practiced.

If the following condition applies, the matrix element rating is 4:

- No gaseous residues result from the service activities.

If neither of the preceding ratings is assigned, complete the checklist below. Assign a rating of 1, 2, or 3 depending on the degree to which the facility meets DFE preferences for this matrix element.

- Are service activities designed to generate minimal and nontoxic gaseous residues in use?
- If CFCs or HCFCs are used in the service activities, have alternatives been thoroughly investigated?
- Are greenhouse gases used or generated in providing the service?
- Are products designed to generate minimal and nontoxic gaseous residues in recycling or disposal?
- Are gaseous residues from service activities recycled?

SERVICE MATRIX ELEMENT: 4,1

Life Stage: Facility Operations
Environmental Stressor: Ecological Impacts

If the following condition applies, the matrix element rating is 0:

- None of the site is left in a natural state and fertilizers and pesticides are used freely.

If the following condition applies, the matrix element rating is 4:

- Essentially all the site has been left in its natural state except for immediate building areas, and no pesticides or fertilizer are used.

If neither of the preceding ratings is assigned, complete the checklist below. Assign a rating of 1, 2, or 3 depending on the degree to which the facility meets DFE preferences for this matrix element.

- Is the maximum possible portion of the facility allowed to remain in its natural state?
- Is the use of pesticides and herbicides on the property minimized?
- Is noise pollution from the site minimized?

SERVICE MATRIX ELEMENT: 4,2

Life Stage: Facility Operations
Environmental Stressor: Energy Use

If the following condition applies, the matrix element rating is 0:

- No site energy management of any kind is practiced.

If the following condition applies, the matrix element rating is 4:

- Energy use for facility operations is minimal and is provided by renewable resources (solar, wind power, and the like).

If neither of the preceding ratings is assigned, complete the checklist below. Assign a rating of 1, 2, or 3 depending on the degree to which the facility meets DFE preferences for this matrix element.

- Is the energy needed for heating and cooling the buildings minimized?
- Is the energy needed for lighting the buildings minimized?
- Is energy efficiency a consideration when buying or leasing facility equipment such as copiers, computers, fan motors, and so on?

SERVICE MATRIX ELEMENT: 4,3

Life Stage: Facility Operations
Environmental Stressor: Solid Residues

If the following condition applies, the matrix element rating is 0:

- Solid residues from facility operations are large and are given the minimum treatment necessary to comply with legal requirements.

If the following condition applies, the matrix element rating is 4:

- Solid residues from facility operations are negligible or are completely recycled on site.

If neither of the preceding ratings is assigned, complete the checklist below. Assign a rating of 1, 2, or 3 depending on the degree to which the facility meets DFE preferences for this matrix element.

- Is the facility designed to minimize the co-mingling of solid waste streams?
- Are solid residues from facility operations reused or recycled to the extent possible?
- Are unusable solid residues from facility operations (including food service) disposed of in an environmentally responsible manner?

SERVICE MATRIX ELEMENT: 4,4

Life Stage: Facility Operations
Environmental Stressor: Liquid Residues

If the following condition applies, the matrix element rating is 0:

- Liquid residues from facility operations are large and are given the minimum treatment necessary to comply with legal requirements.

If the following condition applies, the matrix element rating is 4:

- Liquid residues (including stormwater) from facility operations are negligible or are completely recycled on site.

If neither of the preceding ratings is assigned, complete the checklist below. Assign a rating of 1, 2, or 3 depending on the degree to which the facility meets DFE preferences for this matrix element.

- Is the facility designed to minimize the co-mingling of liquid waste streams?

- Are liquid treatment plants monitored to ensure that they operate at peak efficiency?
- Are unusable liquid residues from facility operations disposed of in an environmentally responsible manner?

SERVICE MATRIX ELEMENT: 4,5

Life Stage: Facility Operations
Environmental Stressor: Gaseous Residues

If the following condition applies, the matrix element rating is 0:

- Gaseous residues from facility operations are large and are given the minimum treatment necessary to comply with legal requirements.

If the following condition applies, the matrix element rating is 4:

- Gaseous residues from facility operations are negligible or are completely recycled on site.

If neither of the preceding ratings is assigned, complete the checklist below. Assign a rating of 1, 2, or 3 depending on the degree to which the facility meets DFE preferences for this matrix element.

- Is facility operations-related transportation to and from the facility minimized?
- Are furnaces, incinerators, and other combustion processes and their related air pollution control devices monitored to ensure operation at peak efficiency?
- Is employee commuting minimized by job sharing, telecommuting, and similar programs?

SERVICE MATRIX ELEMENT: 5,1

Life Stage: Facility and Service Closure
Environmental Stressor: Ecological Impacts

If the following condition applies, the matrix element rating is 0:

- Structures must be completely demolished, with major impacts on natural areas or existing ecosystems.

If the following condition applies, the matrix element rating is 4:

- Site and structures can be reused with negligible impact on natural areas or existing ecosystems.

If neither of the preceding ratings is assigned, complete the checklist below. Assign a rating of 1, 2, or 3 depending on the degree to which the facility meets DFE preferences for this matrix element.

- Will site and system closure have substantial impacts on natural areas?
- Will site and system closure have substantial impacts on existing ecosystems?

SERVICE MATRIX ELEMENT: 5,2

Life Stage: Facility and Service Closure
Environmental Stressor: Energy Use

If the following condition applies, the matrix element rating is 0:

- Very large energy consumption will be required for service and facility closure.

If the following condition applies, the matrix element rating is 4:

- Negligible energy consumption will be required for service and facility closure.

If neither of the preceding ratings is assigned, complete the checklist below. Assign a rating of 1, 2, or 3 depending on the degree to which the facility meets DFE preferences for this matrix element.

- Will site closure require the transport of large quantities of materials?
- Will building demolition or renovation require large energy expenditures?
- Will equipment demolition or renovation require large energy expenditures?

SERVICE MATRIX ELEMENT: 5,3

Life Stage: Facility and Service Closure
Environmental Stressor: Solid Residues

If the following condition applies, the matrix element rating is 0:

- Large amounts of solid residues will be produced by facility and service closure and avoidance is difficult or impossible.

If the following condition applies, the matrix element rating is 4:

- Negligible amounts of solid residues, and none with biotoxic characteristics, will be produced by facility and service closure.

If neither of the preceding ratings is assigned, complete the checklist below. Assign a rating of 1, 2, or 3 depending on the degree to which the facility meets DFE preferences for this matrix element.

- Does the facility contain asbestos or lead paint?
- Can service equipment be reused or recycled?
- Can unwanted building fixtures and components be reused or recycled?
- Can building structural components be reused or recycled?
- Can spare parts, supplies, or other solid service activity consumables or residues be used or recycled?

SERVICE MATRIX ELEMENT: 5,4

Life Stage: Facility and Service Closure
Environmental Stressor: Liquid Residues

If the following condition applies, the matrix element rating is 0:

- Large amounts of liquid residues will be produced by facility and service closure and avoidance is difficult or impossible.

If the following condition applies, the matrix element rating is 4:

- Negligible amounts of liquid residues will be produced by facility and service closure.

If neither of the preceding ratings is assigned, complete the checklist below. Assign a rating of 1, 2, or 3 depending on the degree to which the facility meets DFE preferences for this matrix element.

- Will liquid residues be produced during demolition?
- Will liquid residues be produced during services termination?
- Can liquid service activity consumables or residues be used or recycled?

SERVICE MATRIX ELEMENT: 5,5

Life Stage: Facility and Service Closure
Environmental Stressor: Gaseous Residues

If the following condition applies, the matrix element rating is 0:

- Large amounts of gaseous residues will be produced by facility and service closure and avoidance is difficult or impossible.

If the following condition applies, the matrix element rating is 4:

- Negligible amounts of gaseous residues will be produced by facility and service closure.

If neither of the preceding ratings is assigned, complete the checklist below. Assign a rating of 1, 2, or 3 depending on the degree to which the facility meets DFE preferences for this matrix element.

- Will gaseous residues be produced during demolition?
- Will gaseous residues be produced during services termination?
- Can gaseous service activity consumables or residues be used or recycled?

APPENDIX E

Environmentally Responsible Infrastructure Matrix: Scoring Guidelines and Protocols

The Infrastructure Improvement Matrix is described in Chapter 12. In this appendix, a sample of possible scoring considerations appropriate to each of the matrix elements is presented. It is anticipated that different types of infrastructures may require different check lists and evaluations, so this appendix is presented as an example rather than as a universal formula.

INFRASTRUCTURE MATRIX ELEMENT: 1,1

Life Stage: Site Development
Environmental Stressor: Ecological Impacts/Materials Choice

If the following condition applies, the matrix element rating is 0:

- The infrastructure is developed with massive disturbance or destruction of natural areas.

If the following condition applies, the matrix element rating is 4:

- The infrastructure is developed with negligible destruction of natural areas.

If neither of the preceding ratings is assigned, complete the checklist below. Assign a rating of 1, 2, or 3 depending on the degree to which the infrastructure meets DFE preferences for this matrix element.

- Has the proposed infrastructure site previously been used for similar activities? If not, have any such sites been surveyed for availability?
- Is necessary development activity, if any, planned to avoid disruption of existing biological communities?
- Is the use of heavy equipment for infrastructure site development minimized?
- Are areas disturbed during infrastructure site development carefully restored?

- Is the biota of the infrastructure site compatible with all planned infrastructure activities?

INFRASTRUCTURE MATRIX ELEMENT: 1,2

Life Stage: Site Development
Environmental Stressor: Energy Use

If the following condition applies, the matrix element rating is 0:

- A complete new energy infrastructure is installed during infrastructure development.

If the following condition applies, the matrix element rating is:

- The existing energy infrastructure is used without modification.

If neither of the preceding ratings is assigned, complete the checklist below. Assign a rating of 1, 2, or 3 depending on the degree to which the infrastructure meets DFE preferences for this matrix element.

- Is the infrastructure site such that it can be made operational with only minimal energy expenditures?
- Has the infrastructure site been selected so as to avoid any energy emission impacts on existing biota?
- Does the infrastructure site enable delivery and installation of construction or renovation materials with minimal use of energy?

INFRASTRUCTURE MATRIX ELEMENT: 1,3

Life Stage: Site Development
Environmental Stressor: Solid Residues

If the following condition applies, the matrix element rating is 0:

- A large quantity of solid residues is generated during infrastructure site development and no recycling is practiced.

If the following condition applies, the matrix element rating:

- Insignificant solid residues are generated during infrastructure site development.

If neither of the preceding ratings is assigned, complete the checklist below. Assign a rating of 1, 2, or 3 depending on the degree to which the infrastructure meets DFE preferences for this matrix element.

- Is the infrastructure site such that it can be made operational with only minimal production of solid residues?

- Have plans been made to ensure that any solid residues generated in the process of developing the infrastructure site are managed so as to minimize their impacts on biota and human health?
- If any solid residues generated in the process of developing the infrastructure site may be hazardous or toxic to biota or humans, have plans been made to minimize releases and exposures?

INFRASTRUCTURE MATRIX ELEMENT: 1,4

Life Stage: Site Development
Environmental Stressor: Liquid Residues

If the following condition applies, the matrix element rating is 0:

- A large quantity of liquid residues is generated during infrastructure site development and no recycling is practiced.

If the following condition applies, the matrix element rating is 4:

- Insignificant liquid residues are generated during infrastructure site development.

If neither of the preceding ratings is assigned, complete the checklist below. Assign a rating of 1, 2, or 3 depending on the degree to which the infrastructure meets DFE preferences for this matrix element.

- Is the infrastructure site such that it can be made operational with only minimal production of liquid residues?
- Have plans been made to ensure that any liquid residues generated in the process of developing the infrastructure site are managed so as to minimize their impacts on biota and human health?
- If any liquid residues generated in the process of developing the infrastructure site may be hazardous or toxic to biota or humans, have plans been made to minimize releases and exposures?

INFRASTRUCTURE MATRIX ELEMENT: 1,5

Life Stage: Site Development
Environmental Stressor: Gaseous Residues

If the following condition applies, the matrix element rating is 0:

- A large quantity of gaseous residues is generated during infrastructure site development and no recycling is practiced.

If the following condition applies, the matrix element rating is 4:

- Insignificant gaseous residues are generated during infrastructure site development.

If neither of the preceding ratings is assigned, complete the checklist below. Assign a rating of 1, 2, or 3 depending on the degree to which the infrastructure meets DFE preferences for this matrix element.

- Is the infrastructure site such that it can be made operational with only minimal production of gaseous residues.
- Have plans been made to ensure that any gaseous residues generated in the process of developing the infrastructure site are managed so as to minimize their impacts on biota and human health?
- If any gaseous residues generated in the process of developing the infrastructure site may be hazardous or toxic to biota or humans, have plans been made to minimize releases and exposures?

INFRASTRUCTURE MATRIX ELEMENT: 2,1

Life Stage: Materials and Product Delivery
Environmental Stressor: Ecological Impacts/Materials Choice

If either of the following conditions apply, the matrix element rating is 0:

- For the case where supplier components/subsystems are used: No/little information is known about the chemical content in supplied products and components.
- For the case where materials are acquired from suppliers: A scarce material is used where a reasonable alternative is available. (Scarce materials are defined as antimony, beryllium, boron, cobalt, chromium, gold, mercury, the platinum metals [Pt, Ir, Os, Pa, Rh, Ru], silver, thorium, and uranium.)

If the following condition applies, the matrix element rating is 4:

- No virgin material is used in incoming components or materials.

If neither of the preceding ratings is assigned, complete the checklist below. Assign a rating of 1, 2, or 3 depending on the degree to which the infrastructure meets DFE preferences for this matrix element.

- Is the infrastructure designed to minimize the use of materials in restricted supply (see preceding list)?
- Is the infrastructure designed to utilize recycled materials or components wherever possible?

INFRASTRUCTURE MATRIX ELEMENT: 2,2

Life Stage: Materials and Product Delivery
Environmental Stressor: Energy Use

If the following condition applies, the matrix element rating is 0:

- One or more of the principal materials used in the infrastructure requires energy-intensive extraction and suitable alternative materials are available that do not. (Materials requiring energy-intensive extraction are defined as virgin aluminum, virgin steel, and virgin petroleum.)

If the following condition applies, the matrix element rating is 4:

- Negligible energy is needed to extract or ship the materials or components for this infrastructure.

If neither of the preceding ratings is assigned, complete the checklist below. Assign a rating of 1, 2, or 3 depending on the degree to which the infrastructure meets DFE preferences for this matrix element.

- Is the infrastructure designed to minimize the use of virgin materials whose extraction is energy intensive?
- Does the infrastructure design avoid or minimize the use of high-density materials whose transport to and from the infrastructure site will require significant energy use? (Such materials are defined as those with a specific gravity above 7.0.)
- Is transport distance of incoming materials and components minimized?

INFRASTRUCTURE MATRIX ELEMENT: 2,3

Life Stage: Materials and Product Delivery
Environmental Stressor: Solid Residues

If the following condition applies, the matrix element rating is 0:

- Solid residues from shipping and installation are large and no reuse/recycling programs are in use.

If the following condition applies, the matrix element rating is 4:

- Solid residues from shipping and installation are minor and are >90 percent reused/recycled.

If neither of the preceding ratings is assigned, complete the checklist below. Assign a rating of 1, 2, or 3 depending on the degree to which the infrastructure meets DFE preferences for this matrix element.

- Have solid shipping and installation residues been minimized to the greatest extent possible?
- Has the resale of all solid shipping and installation residues been implemented?

INFRASTRUCTURE MATRIX ELEMENT: 2,4

Life Stage: Materials and Product Delivery
Environmental Stressor: Liquid Residues

If the following condition applies, the matrix element rating is 0:

- Liquid shipping and installation residues are large and no reuse/recycling programs are in use.

If the following condition applies, the matrix element rating is 4:

- Liquid shipping and installation residues are minor and each constituent is >90 percent reused/recycled.

If neither of the preceding ratings is assigned, complete the checklist below. Assign a rating of 1, 2, or 3 depending on the degree to which the infrastructure DFE preferences for this matrix element.

- If solvents or oils are used in the shipping and installation of this infrastructure, is their use minimized and have alternatives been investigated and implemented?
- If water is used in the shipping and installation of this infrastructure, is its use minimized and have alternative approaches been investigated and implemented?

INFRASTRUCTURE MATRIX ELEMENT: 2,5

Life Stage: Materials and Product Delivery
Environmental Stressor: Gaseous Residues

If either of the following conditions apply, the matrix element rating is 0:

- Gaseous infrastructure shipping and installation residues are large and are uncontrolled.
- CFCs are used in infrastructure shipping and installation.

If the following condition applies, the matrix element rating is 4:

- Gaseous infrastructure shipping and installation residues are relatively minor and reuse/recycling programs are in use.

If neither of the preceding ratings is assigned, complete the checklist below. Assign a rating of 1, 2, or 3 depending on the degree to which the infrastructure DFE preferences for this matrix element.

- Are greenhouse gases used or generated in any shipping or installation activity connected with the infrastructure? Have the emitted amounts been minimized?

INFRASTRUCTURE MATRIX ELEMENT: 3,1

Life Stage: Infrastructure Manufacture
Environmental Stressor: Ecological Impacts/Materials Choice

If the following condition applies, the matrix element rating is 0:

- Infrastructure manufacture requires relatively large amounts of materials that are restricted, toxic, and/or radioactive.

If the following condition applies, the matrix element rating is 4:

- Materials used in infrastructure manufacture are completely closed loop (captured and reused/recycled) with minimum inputs required.

If neither of the preceding ratings is assigned, complete the checklist below. Assign a rating of 1, 2, or 3 depending on the degree to which the infrastructure manufacture meets DFE preferences for this matrix element.

- Does the infrastructure design avoid the use of materials that are in restricted supply?
- Is the use of toxic material avoided or minimized?
- Is the use of radioactive material avoided?
- Is the use of virgin material minimized?
- Has the chemical treatment of materials and components been minimized?

INFRASTRUCTURE MATRIX ELEMENT: 3,2

Life Stage: Infrastructure Manufacture
Environmental Stressor: Energy Use

If the following condition applies, the matrix element rating is 0:

- Energy use for infrastructure manufacture and installation is high and less energy-intensive alternatives are available.

If the following condition applies, the matrix element rating is 4:

- Process equipment manufacture and installation requires no or minimal energy use.

If neither of the preceding ratings is assigned, complete the checklist below. Assign a rating of 1, 2, or 3 depending on the degree to which the infrastructure meets DFE preferences for this matrix element.

- Is the infrastructure manufacture designed to minimize the use of energy-intensive processing steps?
- Is the infrastructure manufacture designed to minimize energy-intensive evaluation/testing steps?
- Does the infrastructure manufacture use co-generation, heat exchanges, and/or other techniques to utilize otherwise wasted energy?

INFRASTRUCTURE MATRIX ELEMENT: 3,3

Life Stage: Infrastructure Manufacture
Environmental Stressor: Solid Residues

If either of the following conditions apply, the matrix element rating is 0:

- For the case where materials are acquired from suppliers: Metals from virgin ores are used, creating substantial waste rock residues that could be avoided by the use of virgin material, and suitable virgin material is available from recycling streams.
- For the case where supplier components/subsystems are used: All incoming packaging is from virgin sources and consists of three or more types of materials.

If all of the following conditions apply, the matrix element rating is 4:

- For the case where materials are acquired from suppliers: No solid residues result from resource extraction or during production of materials by recycling (example: petroleum).
- For the case where supplier components/sub-systems are used: Minimal or no packaging material is used or supplier takes back all packaging material.
- For the case where supplier components/subsystems are used: Incoming packaging is totally reused/recycled.

If neither of the preceding ratings is assigned, complete the checklist below. Assign a rating of 1, 2, or 3 depending on the degree to which the infrastructure meets DFE preferences for this matrix element.

- Is the infrastructure designed to minimize the use of materials whose extraction or purification involves the production of large amounts of solid residues (i.e., coal and all virgin metals)?
- Is the infrastructure designed to minimize the use of materials whose extraction or purification involves the production of toxic solid residues? (This category includes all radioactive materials.)
- Has incoming packaging volume and weight, at and among all levels (primary, secondary, and tertiary), been minimized?
- Is materials diversity minimized in incoming packaging?

INFRASTRUCTURE MATRIX ELEMENT: 3,4

Life Stage: Infrastructure Manufacture
Environmental Stressor: Liquid Residues

If either of the following conditions apply, the matrix element rating is 0:

- For the case where supplier components/subsystems are used: Metals from virgin ores that cause substantial acid mine drainage are used, and suitable virgin material is available from recycling streams. (Materials causing acid mine drainage are defined as copper, iron, nickel, lead, and zinc.)
- For the case where materials are acquired from suppliers: The packaging contains toxic or hazardous substances that might leak from it if improper disposal occurs.

If both of the following conditions apply, the matrix element rating is 4:

Appendix E   Environmentally Responsible Infrastructure Matrix   299

- For the case where materials are acquired from suppliers: No liquid residues result from resource extraction or during production of materials by recycling.
- For the case where supplier components/subsystems are used: No liquid residue is generated during transportation, unpacking, or use of this infrastructure.

If neither of the preceding ratings is assigned, complete the checklist below. Assign a rating of 1, 2, or 3 depending on the degree to which the infrastructure meets DFE preferences for this matrix element.

- Is the infrastructure designed to minimize the use of materials whose extraction or purification involves the generation of large amounts of liquid residues? (This category includes paper and allied products, coal, and materials from biomass.)
- Is the infrastructure designed to minimize the use of materials whose extraction or purification involves the generation of toxic liquid residues? (These materials are defined as aluminum, copper, iron, lead, nickel, and zinc.)
- Are refillable/reusable containers used for incoming liquid materials where appropriate?
- Does the use of incoming components require cleaning that involves a large amount of water or that generate liquid residues needing special disposal methods?

INFRASTRUCTURE MATRIX ELEMENT: 3,5

Life Stage: Infrastructure Manufacture
Environmental Stressor: Gaseous Residues

If the following condition applies, the matrix element rating is 0:

- The materials used cause substantial emissions of toxic, smog-producing, or greenhouse gases into the environment, and suitable alternatives that do not do so are available. (These materials are defined as aluminum, copper, iron, lead, nickel, zinc, paper and allied products, and concrete.)

If the following condition applies, the matrix element rating is 4:

- No gaseous residues are produced during resource extraction or production of materials by recycling.

If neither of the preceding ratings is assigned, complete the checklist below. Assign a rating of 1, 2, or 3 depending on the degree to which the infrastructure meets DFE preferences for this matrix element.

- Is the infrastructure designed to minimize the use of materials whose extraction or purification involves the generation of large amounts of gaseous (toxic or otherwise) residues? (Such materials are defined as aluminum, copper, iron, lead, nickel, and zinc.)

INFRASTRUCTURE MATRIX ELEMENT: 4,1

Life Stage: Infrastructure Use
Environmental Stressor: Ecological Impacts/Materials Choice

If the following condition applies, the matrix element rating is 0:

- Consumables contain significant quantities of materials in restricted supply or toxic/hazardous substances.

If the following condition applies, the matrix element rating is 4:

- Infrastructure use and infrastructure maintenance requires no consumables.

If neither of the preceding ratings is assigned, complete the checklist below. Assign a rating of 1, 2, or 3 depending on the degree to which the infrastructure meets DFE preferences for this matrix element.

- Has consumable material use been minimized?
- Have the materials been chosen such that no environmentally inappropriate maintenance is required, and no unintentional release of toxic materials to the environment occurs during use?
- Are consumable materials generated from recycled streams rather than virgin material?

INFRASTRUCTURE MATRIX ELEMENT: 4,2

Life Stage: Infrastructure Use
Environmental Stressor: Energy Use

If the following condition applies, the matrix element rating is 0:

- Infrastructure use and/or maintenance is relatively energy intensive and less energy-intensive methods are available to accomplish the same purpose.

If the following condition applies, the matrix element rating is 4:

- Infrastructure use and maintenance requires little or no energy.

If neither of the preceding ratings is assigned, complete the checklist below. Assign a rating of 1, 2, or 3 depending on the degree to which the infrastructure meets DFE preferences for this matrix element.

- Has the infrastructure been designed to minimize energy use while in service?
- Has energy use during maintenance/repair been minimized?
- Have energy-conserving design features (such as auto shut off or enhanced insulation) been incorporated?
- Can the infrastructure monitor and display its energy use and/or its operating energy efficiency while in service?

INFRASTRUCTURE MATRIX ELEMENT: 4,3

Life Stage: Infrastructure Use
Environmental Stressor: Solid Residues

If the following condition applies, the matrix element rating is 0:

- Infrastructure generates significant quantities of hazardous/toxic solid residue during use or from repair/maintenance operations.

If the following condition applies, the matrix element rating is 4:

- Infrastructure generates no (or relative minor amounts of) solid residue during use or from repair/maintenance operations.

If neither of the preceding ratings is assigned, complete the checklist below. Assign a rating of 1, 2, or 3 depending on the degree to which the infrastructure meets DFE preferences for this matrix element.

- Has the periodic disposal of solid materials (such as cartridges, containers, or batteries) associated with the use and/or maintenance of this infrastructure been avoided or minimized?
- Is the infrastructure equipment modular so that it can be easily repaired and upgraded?
- Have alternatives to the use of solid consumables been thoroughly investigated and implemented where appropriate?
- If intentional dissipative emissions to land occur as a result of using this infrastructure, have less environmentally harmful alternatives been investigated?

INFRASTRUCTURE MATRIX ELEMENT: 4,4

Life Stage: Infrastructure Use
Environmental Stressor: Liquid Residues

If the following condition applies, the matrix element rating is 0:

- Infrastructure generates significant quantities of hazardous/toxic liquid residue during use or from repair/maintenance operations.

If the following condition applies, the matrix element rating is 4:

- Infrastructure generates no (or relatively minor amounts of) liquid residues during use or from repair/maintenance operations.

If neither of the preceding ratings is assigned, complete the checklist below. Assign a rating of 1, 2, or 3 depending on the degree to which the infrastructure meets DFE preferences for this matrix element.

- Has the periodic disposal of liquid materials (such as lubricants and hydraulic fluids) associated with the use and/or maintenance of this infrastructure been avoided or minimized?

- Have alternatives to the use of liquid consumables been thoroughly investigated and implemented where appropriate?
- If intentional dissipative emissions to water occur as a result of using this infrastructure, have less environmentally harmful alternatives been investigated?
- If infrastructure contains liquid material that has the potential to be unintentionally dissipated during use or repair, have appropriate preventive measures been incorporated?

INFRASTRUCTURE MATRIX ELEMENT: 4,5

Life Stage: Infrastructure Use
Environmental Stressor: Gaseous Residues

If the following condition applies, the matrix element rating is 0:

- Infrastructure generates significant quantities of hazardous/toxic gaseous residue during use or from repair/maintenance operations.

If the following condition applies, the matrix element rating is 4:

- Infrastructure generates no (or relatively minor amounts of) gaseous residues during use or from repair/maintenance operations.

If neither of the preceding ratings is assigned, complete the checklist below. Assign a rating of 1, 2, or 3 depending on the degree to which the infrastructure meets DFE preferences for this matrix element.

- Has the periodic emission of gaseous materials (such as CO_2, SO_2, VOCs, and CFCs) associated with the use and/or maintenance of this infrastructure been avoided or minimized?
- Have alternatives to the use of gaseous consumables been thoroughly investigated and implemented where appropriate?
- If intentional dissipative emissions to air occur as a result of using or maintaining this infrastructure, have less environmentally harmful alternatives been investigated?
- If infrastructure contains any gaseous materials that have the potential to be unintentionally dissipated during use or repair, have the appropriate preventive measures been incorporated?

INFRASTRUCTURE MATRIX ELEMENT: 5,1

Life Stage: Refurbishment, Recycling, Disposal
Environmental Stressor: Ecological Impacts/Materials Choice

If the following condition applies, the matrix element rating is 0:

- Structures must be completely demolished, with major impacts on natural areas or existing ecosystems.

If the following condition applies, the matrix element rating is 4:

- Site and structures can be reused with negligible impact on natural areas or existing ecosystems.

If neither of the preceding ratings is assigned, complete the checklist below. Assign a rating of 1, 2, or 3 depending on the degree to which the infrastructure meets DFE preferences for this matrix element.

- Will infrastructure site closure have substantial impacts on natural areas?
- Will infrastructure site closure have substantial impacts on existing ecosystems?

INFRASTRUCTURE MATRIX ELEMENT: 5,2

Life Stage: Refurbishment, Recycling, Disposal
Environmental Stressor: Energy Use

If the following condition applies, the matrix element rating is 0:

- Very large energy consumption will be required for infrastructure closure.

If the following condition applies, the matrix element rating is 4:

- Negligible energy consumption will be required for infrastructure closure.

If neither of the preceding ratings is assigned, complete the checklist below. Assign a rating of 1, 2, or 3 depending on the degree to which the infrastructure meets DFE preferences for this matrix element.

- Will infrastructure site closure require the transport of large quantities of materials?
- Will building demolition or renovation require large energy expenditures?
- Will component removal or renovation require large energy expenditures?

INFRASTRUCTURE MATRIX ELEMENT: 5,3

Life Stage: Refurbishment, Recycling, Disposal
Environmental Stressor: Solid Residues

If the following condition applies, the matrix element rating is 0:

- Large amounts of solid residues will be produced by infrastructure closure and avoidance is difficult or impossible.

If the following condition applies, the matrix element rating is 4:

- Negligible amounts of solid residues, and none with biotoxic characteristics, will be produced by infrastructure closure.

If neither of the preceding ratings is assigned, complete the checklist below. Assign a rating of 1, 2, or 3 depending on the degree to which the infrastructure meets DFE preferences for this matrix element.

- Does the infrastructure contain asbestos or lead paint?
- Can infrastructure equipment be reused or recycled?
- Can unwanted building fixtures and components be reused or recycled?
- Can building structural components be reused or recycled?
- Can spare parts, supplies, or other solid infrastructure activity consumables or residues be used or recycled?

INFRASTRUCTURE MATRIX ELEMENT: 5,4

Life Stage: Refurbishment, Recycling, Disposal
Environmental Stressor: Liquid Residues

If the following condition applies, the matrix element rating is 0:

- Large amounts of liquid residues will be produced by infrastructure closure and avoidance is difficult or impossible.

If the following condition applies, the matrix element rating is 4:

- Negligible amounts of liquid residues will be produced by infrastructure closure.

If neither of the preceding ratings is assigned, complete the checklist below. Assign a rating of 1, 2, or 3 depending on the degree to which the infrastructure meets DFE preferences for this matrix element.

- Will liquid residues be produced during demolition?
- Will liquid residues be produced during infrastructure termination?
- Can liquid infrastructure activity consumables or residues be used or recycled?

INFRASTRUCTURE MATRIX ELEMENT: 5,5

Life Stage: Refurbishment, Recycling, Disposal
Environmental Stressor: Gaseous Residues

If the following condition applies, the matrix element rating is 0:

- Large amounts of gaseous residues will be produced by infrastructure closure and avoidance is difficult or impossible.

If the following condition applies, the matrix element rating is 4:

- Negligible amounts of gaseous residues will be produced by infrastructure closure.

If neither of the preceding ratings is assigned, complete the checklist below. Assign a rating of 1, 2, or 3 depending on the degree to which the infrastructure meets DFE preferences for this matrix element.

- Will gaseous residues be produced during demolition?
- Will gaseous residues be produced during termination of infrastructure operations?
- Can gaseous infrastructure consumables or residues be used or recycled?

APPENDIX F

Values of Localization Parameters for Common Stressors

In chapter 13, a procedure was developed for localization of the matrix assessment in which stressor impacts on time (t), space (d), peril (p), and exposure (e) are explicitly taken into account. This appendix provides time and distance parameter values for common stressors. The parameters for peril and exposure must be derived with local and regional conditions taken into account.

Predominant Stressor	Receptor Group	$t_{i,j}$	$d_{i,j}$
CFC-113	GG	0	0
CO_2	E	0	0
Coal	E	0	0
NO_x	GR	3	2
Oil	LL	2	4
Oil	LR	2	4
Oil	RC	0	0
Ore	SR	1	3
Packaging	RC	2	3
Packaging	SR	2	3
Scrap	SR	1	3
SO_2	GR	2	1
Solvents	LR	2	3
Steel	MR	1	2
Steel	RC	1	2
Tires	SR	1	3
VOC	GL	3	2
VOC	GR	3	2

Receptor groups are defined as follows:
RC, resource consumption; B, biodiversity; E, energy use; W, water use; SR, solid residue generation; LL, liquid residue generation, local impacts; LR, liquid residue generation, regional impacts; GL, gaseous residue generation, local impacts; GR, gaseous residue generation, regional impacts; GG, gaseous residue generation, global impacts; MR, materials recycling.

Index

Acid deposition, 6, 8, 10
Aesthetic richness, 6, 8, 10
Allenby, Braden R., 100
Allocation, 39–41
Alpha services, 151
American Fiber Manufacturers Association, 76, 225
Asnaes Power Company, 176
Association of the Dutch Chemical Industry (*see* Netherlands VNCI System)
AT&T Corporation, 100, 170
Atmospheric impacts, 47–48
Atom utilization concept, 53
Automobile,
    front ends, life-cycle assessment, 55–57
    generic life-cycle assessment, 103–108
    manufacturing facilities, life-cycle assessment, 142–147, 201–203
    manufacturing processes, life-cycle assessment, 118–132
    repair, life-cycle assessment, 156–165
Automotive infrastructure, 185–192
Ayres, Robert, 26

Bank transactions, life-cycle assessment, 165–167
Battelle Corporation, 84, 93–95
Beta services, 151–153
Biodiversity, 6, 8, 10
Bonus scoring, 193–194
Boundaries, 30–33, 69–70
    level of detail, 31
    life-stage, 30
    natural ecosystem, 31–33
    space and time, 32

Brent Spar Oil Platform, 115
British Research Establishment, 140
Brundtland Commission (*see* World Commission on Environment and Development)
Budgets, product and process, 36–39
Building materials, life-cycle assessment, 84

Characterization, 46
Classification, 46
Climate change (*see* Global climate change)
Complementary process, 115, 117–118, 124–125
Computer workstations, life-cycle assessment, 80–83
Contracts, buildings and equipment, 156
Corporate life-cycle assessment, 173–175
Council for Solid Waste Solutions, 78
Cradle-to-warehouse, 88

Data limitations, 70–71
Diapers, life-cycle assessment, 34, 83–84
Die casting, 120–121
Dow Chemical Company, 90–92

Ecoindustrial parks (*see* Ecoparks)
Ecoparks, 175
Ecoscreening, 87–88
Elements, supply status, 197
Energy Star Program, 83
Environmental audit, 231
Environmental impact assessment, 231
Environmental load unit (ELU), 53–55
EPA Science Advisory Board, 9, 195
EPS System, 52–57, 75

Facility,
    audits, 147
    life-cycle assessment, 134–149
    life-cycle stages, 135–139
    operations, 138
    scoring guidelines and protocols, 265–277
Field, Frank, 14–15
Five percent rule, 31
Food products, life-cycle assessment, 79–80
Fossil fuels, depletion, 6, 8, 10
Franklin Associates, 41, 76, 225
Functional unit, 34–36, 70

Gamma services, 153
Gate-to-gate, 88
Global climate change, 6, 8, 10
Goal setting (*see* Life-cycle assessment, goal and scope)
Grand objectives, 3, 207–213
Green building design reviews, 147
Green Buildings Council, 139
Greenpeace, 115
Grocery sacks, life-cycle assessment, 78
Gyproc Corporation, 176

Hocking, Martin, 33
Hot-drink cups, life-cycle assessment, 34
Hot-drink machines, life-cycle assessment, 78–79
Human organism damage, 6, 8, 10
Human species extinction, 6

IBM Corporation, 90
Impact analysis (*see* Life-cycle assessment, impact analysis)
Impacts,
    atmosphere, 47–48
    soil, 49–50
    water, 49–50
Infrastructure,
    life-cycle assessment, 182–192
    life-cycle stages, 183–184
    scoring guidelines and protocols, 291–304

Intercorporate life-cycle assessment, 175–178
International Geosphere-Biosphere Program (IGBP), 75
Interpretation analysis (*see* Life-cycle assessment, interpretation analysis)
Inventory analysis (*see* Life-cycle assessment, inventory analysis)
Inventory flow diagram, 33, 35

Jacobs Engineering, 95

Kalundborg, Denmark, 176–178

Land use, 6, 8, 10
Landfill exhaustion, 6, 8, 11
Life-cycle assessment, 18
    case studies, 34, 55–57, 58–62, 76–84
    corporate, 173–175
    facility, 134–149
    goal and scope, 21–23, 29–33
    infrastructure, 182–192
    impact analysis, 23, 45–61, 71–75
    intercorporate, 175–178
    interpretation analysis, 23–24, 62–66
    inventory analysis, 23, 34–41
    process, 111–132
    product, 99–109
    service, 150–168
Life-cycle stages,
    facility, 135–149
    infrastructure, 183–184
    process, 112–115
    product, 18–19
    service, 151–153
Lithographic printing, life-cycle assessment, 84
Localization, 46, 195–203, 305
Low-hanging fruit, 4

Matrix,
    calculations, 96–97
    concepts, 100–102
    weighting, by consensus, 203–207
    weighting, by grand objectives, 207–213
Metal cleaning, 124–125

Metal forming, 121–122
Metal plating, 123
Microelectronics and Computer Technology Corporation (MCC), 80
Migros Corporation, 90
Monsanto Company, 92
Montreal Protocol, 5
Motorola Corporation, 92–93
Multi-objective decision making, 13–15

Natural Step, 7
Needs-based approaches, 223–227
Netherlands,
    NSAEL Method, 58–62, 75
    VNCI System, 49–52, 75
Nonlinearities, 71–73
Novo Nordisk Corporation, 176

Odor, 6, 11
Oil spills, 6, 11
Ozone depletion, 6, 8, 10

Painting, 123
Paper manufacture, inventory flow diagram, 33
Pareto plot, 64, 132
Patagonia, Inc., 76
Polyester blouses, life-cycle assessment, 76–78
Portable radio, inventory flow diagram, 35
Primary process, 114
Prioritization, 62–65, 127–132
Process,
    life-cycle assessment, 111–132
    life-cycle stages, 112–115
    scoring guidelines and protocols, 235–249
    technology study, 231
Procter & Gamble Company, 41
Product,
    life-cycle assessment, 99–109
    life-cycle assessment, integrated, 170–173
    life-cycle stages, 18–19
    scoring guidelines and protocols, 235–249
Radionuclides, 8, 11

Refrigerators, life-cycle assessment, 34
Resources, depletion, 6, 8, 10–11
Reverse life-cycle assessment, 216–228
Rio Treaty, 5

Sand casting, 118–120
Scope determination (*see* Life-cycle assessment, goal and scope)
Scoping, 87–88
Scoring guidelines and protocols,
    facility, 265–277
    infrastructure, 291–304
    process, 250–264
    product, 235–249
    service, 278–290
Service,
    economy, 167–168
    life-cycle assessment, 150–168
    life-cycle stages, 151–153
    scoring guidelines and protocols, 278–290
SETAC (Society of Environmental Toxicology and Chemistry), 18
Sheldon, Roger, 52
Shell Oil Corporation, 115
Smog, 6, 8, 10
Soil,
    depletion, 6, 8, 10
    impacts, 49–50
Spatial scales, 74–75
Statoil Corporation, 176
Streamlined life-cycle assessment,
    assets and liabilities, 97
    localized and valuated, 213–215
    philosophy, 87–97
Stressor, 45, 47–49, 305
Surfactants, 41–43
Sustainable development, 4, 6
Swedish Environmental Institute (IVL), 52–57
Swedish Waste Research Council, 79

Target plot,
    automobile manufacturing facilities, 148
    automobile manufacturing processes, 129–130

automobile repair facilities, 164–165
automobiles, generic, 108–109
automotive infrastructure, 191–192
concepts, 102–103
Kalundborg complex, 179–180
localized and valuated, 214
telecommunications products, 172
weighted, 206, 208
Temporal scales, 73–74
Thermal pollution, 6, 8, 11
Thresholds, 71–73
Trimming and smoothing, 124

U.S. Environmental Protection Agency (EPA), 80
U.S. National Institute for Standards and Technology, 140
Ultimate life-cycle assessment, 230–234
University of Amsterdam, 78, 84
University of British Columbia, 90, 140
University of California at Los Angeles (UCLA), 76
Upgraded streamlining, 193–215
Utilization economy, 167

Valuation, 46, 75
Van den Berg, N.W., 231
Volvo Car Company, 52–57

Washing machine, life-cycle assessment, 216–223
Water,
    availability and quality, 6, 8, 10
    impacts, 49–50
Weitz, Keith, 88
Welding, 122
Window frames, life-cycle assessment, 58–62
World Commission on Environment and Development, 4, 7